Gems of Combinatorial Optimization and Graph Algorithms

Andreas S. Schulz · Martin Skutella
Sebastian Stiller · Dorothea Wagner
Editors

Gems of Combinatorial Optimization and Graph Algorithms

Editors
Andreas S. Schulz
Technische Universität München
Munich
Germany

Martin Skutella
Technische Universität Berlin
Berlin
Germany

Sebastian Stiller
Technische Universität Braunschweig
Braunschweig
Germany

Dorothea Wagner
Karlsruher Institut für Technologie
Karlsruhe
Germany

ISBN 978-3-319-24970-4   ISBN 978-3-319-24971-1   (eBook)
DOI 10.1007/978-3-319-24971-1

Library of Congress Control Number: 2015953249

Springer Cham Heidelberg New York Dordrecht London
© Springer International Publishing Switzerland 2015
This work is subject to copyright. All rights are reserved by the Publisher, whether the whole or part of the material is concerned, specifically the rights of translation, reprinting, reuse of illustrations, recitation, broadcasting, reproduction on microfilms or in any other physical way, and transmission or information storage and retrieval, electronic adaptation, computer software, or by similar or dissimilar methodology now known or hereafter developed.
The use of general descriptive names, registered names, trademarks, service marks, etc. in this publication does not imply, even in the absence of a specific statement, that such names are exempt from the relevant protective laws and regulations and therefore free for general use.
The publisher, the authors and the editors are safe to assume that the advice and information in this book are believed to be true and accurate at the date of publication. Neither the publisher nor the authors or the editors give a warranty, express or implied, with respect to the material contained herein or for any errors or omissions that may have been made.

Printed on acid-free paper

Springer International Publishing AG Switzerland is part of Springer Science+Business Media (www.springer.com)

To the joint mentor of the contributors to this book, Prof. Dr. Rolf H. Möhring, on the occasion of his retirement.

Preface

This book is intended as an homage to Prof. Rolf H. Möhring, an accomplished leader in the field of combinatorial optimization and graph algorithms. We contacted his former advisees who have successfully launched their own academic careers. We asked them each to pick a topic that they thought would be both interesting and accessible to a wide audience with a basic knowledge of graphs, algorithms, and optimization. The result is this collection of beautiful results.

The reader can learn about the connection between shortest paths and mechanism design, about the interplay of priority rules in scheduling and the existence of pure-strategy Nash equilibria in weighted congestion games, about the critical role played by matroids in the existence of pure-strategy Nash equilibria in resource-buying games, about some geometric commonality between proportional resource allocation and selfish flows, about using the gasoline puzzle and the adjacency structure of the matching polytope to solve the budgeted matching problem, about the relation between the knotting graph and the linear structure of graphs, about convex programming relaxations and randomized rounding in scheduling, about the significance of motifs in network analysis, about the analogy between contraction hierarchies used for fast shortest path computations and (perfect) elimination orderings in graphs, about universally good algorithms for the knapsack problem with varying capacity and for a scheduling problem, about the pivotal role of Hanan grids for the minimum Steiner tree problem for rectilinear distances, about a linear-time algorithm that computes the longest tour for points in the plane under the taxicab distance, and about a characterization of certain rectangular dissections with surprising applications.

Each chapter can easily be used as the basis for a lecture or two in an advanced undergraduate course or in a graduate course on graph algorithms, combinatorial optimization, algorithmic game theory, or computational geometry. For improved readability, citations within a chapter are kept to a minimum, but each chapter concludes with a discussion of the relevant literature and provides pointers for further reading.

We are indebted to all the contributors for their enthusiasm about our idea and the time and care they put into preparing their chapters. We also wish to thank Martin Peters for his excitement about our proposal and for making this book possible and Ruth Allewelt for all her support and helpful guidance during the various phases of this project. Finally, on behalf of all his former advisees, we would like to express our deep gratitude to Rolf Möhring. Learning from and working with him has sharpened and broadened our minds. The topics covered in this book are reflective of his wide interests. He has been a wonderful mentor and a true source of inspiration in all these years.

Andreas S. Schulz
Martin Skutella
Sebastian Stiller
Dorothea Wagner

Contents

Shifting Segments to Optimality 1
Stefan Felsner

Linear Structure of Graphs and the Knotting Graph 13
Ekkehard Köhler

Finding Longest Geometric Tours 29
Sándor P. Fekete

Generalized Hanan Grids for Geometric Steiner Trees in Uniform Orientation Metrics 37
Matthias Müller-Hannemann

Budgeted Matching via the Gasoline Puzzle 49
Guido Schäfer

Motifs in Networks 59
Karsten Weihe

Graph Fill-In, Elimination Ordering, Nested Dissection and Contraction Hierarchies 69
Ben Strasser and Dorothea Wagner

Shortest Path to Mechanism Design 83
Rudolf Müller and Marc Uetz

Selfish Routing and Proportional Resource Allocation 95
Andreas S. Schulz

Resource Buying Games 103
Tobias Harks and Britta Peis

Linear, Exponential, but Nothing Else 113
Max Klimm

Convex Quadratic Programming in Scheduling 125
Martin Skutella

Robustness and Approximation for Universal Sequencing 133
Nicole Megow

A Short Note on Long Waiting Lists 143
Sebastian Stiller

Shifting Segments to Optimality

Stefan Felsner

Abstract We begin with two problems which do not appear to be related. Then we use the 'air-pressure' method to prove a theorem about rectangular dissections with prescribed rectangle areas. A corollary to this theorem characterizes area-universal rectangular dissections. These dissections happen to be central to the solution of the two problems.

1 Introduction and Two Problems

A rectangular dissection is a partition of a frame rectangle into rectangles, Fig. 1 shows an example. Rectangular dissections are studied in various fields, see Fig. 2.

- Architects look at them in the context of floorplan generation [11, 14].
- Floorplaning is relevant for module placement in VLSI design [4, 21].
- In graph drawing, rectangular dissections play a role in various representation models for planar graphs [8, 12].
- In cartography, rectilinear dissections are studied as a special class of cartograms [13, 17, 19].

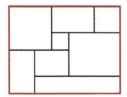

Fig. 1 A rectangular dissection

S. Felsner (✉)
Institut für Mathematik, Technische Universität Berlin, Straße des 17. Juni 136, 10623 Berlin, Germany
e-mail: felsner@math.tu-berlin.de

© Springer International Publishing Switzerland 2015
A.S. Schulz et al. (eds.), *Gems of Combinatorial Optimization and Graph Algorithms*, DOI 10.1007/978-3-319-24971-1_1

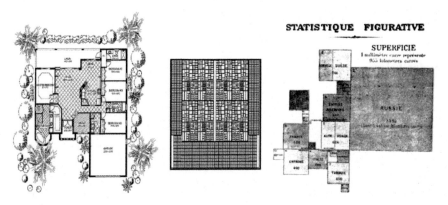

Fig. 2 Near rectangular dissections in applications

In the applications the areas of the rectangles of a dissection are relevant. In many cases these areas are prescribed. A rectangular dissection is *area-universal* if for any assignment of positive weights to the rectangles there is a combinatorially equivalent dissection such that the areas of rectangles are equal to the given weights.

Central to this chapter is the characterization of area-universal rectangular dissections (Theorem 5). In Sect. 2 we state the theorem and discuss some proofs and generalizations. Before getting there we present two problems which do not appear to have much in common. In Sect. 3 we show that both problems can be solved by clever applications of the theorem.

First Problem

By default a *dissection* shall be a rectangular dissection. A dissection is *generic* if it has no cross, i.e., no point where four rectangles of the partition meet. A *segment* of a dissection is a maximal nondegenerate interval that belongs to the union of the boundaries of the rectangles. In general we disregard the four segments from the boundary frame, i.e., we only consider inner segments. Segments are either hori-

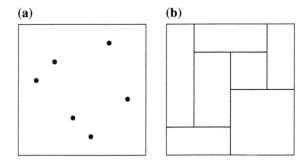

Fig. 3 A generic set of six points and a generic dissection with six segments

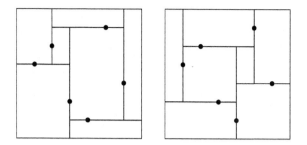

Fig. 4 Two cover maps from the dissection of Fig. 3b to the point set of Fig. 3a

zontal or vertical. The segments of a generic dissection are internally disjoint. Two dissections R and R' are *weakly equivalent* if there exists a bijection ϕ between their segments that preserves the orientation (horizontal/vertical) and such that a segment s has an endpoint on a segment t in R iff $\phi(s)$ has an endpoint on $\phi(t)$ in R'. A set P of points in $\mathbb{R}^2$ is *generic* if no two points from P have the same x or y coordinate, see Fig. 3.

Let P be a set of n points in a rectangular frame F and let R be a generic dissection with n segments. A *cover map* from R to P is a dissection R' that is weakly equivalent to R and has outer rectangle F such that every segment of R' contains exactly one point from P. Figure 4 shows an example.

Problem 1 Does a cover map from R to P exist for all pairs (R, P) where R is a generic dissection with n segments and P is a generic set of n points?

Second Problem

This problem is about *rectilinear duals* of planar graphs. In this drawing model the vertices are represented by simple rectilinear polygons, while edges are represented by side-contacts between the corresponding polygons, see Fig. 5.

Now assume that positive weights $w(v)$ have been assigned to the vertices of the graph. A *rectilinear cartogram* is a rectilinear dual with the additional property that for all vertices the area of the polygon representing v equals $w(v)$. A relevant parameter measuring the complexity of a cartogram is the maximum number of sides of any polygon.

Problem 2 What is the minimum number k such that any given planar triangulation, i.e., maximally planar graph, with positive weights $w(v)$ admits a rectilinear cartogram with $\leq k$-sided polygons in a rectangular frame F of area $\sum_v w(v)$?

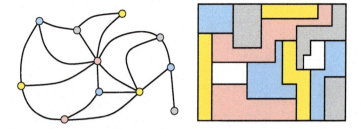

Fig. 5 A graph with a rectilinear dual containing two holes, the *white* regions

2 Area-Universality of Weak Equivalence Classes

The following theorem is a formalization of the statement given in the heading: *weak equivalence classes are area-universal.*

Theorem 3 *Let R be a dissection with rectangles $r_1, \ldots, r_{n+1}$, let the frame F be a rectangle and let $w : \{1, \ldots, n+1\} \to \mathbb{R}_+$ be a weight function with $\sum_i w(i) =$ area(F). There exists a unique dissection R' contained in F that is weakly equivalent to R such that the area of the rectangle r'_i in R' is $w(i)$.*

We will discuss a proof based on an iterative approach below. Before getting there, however, we introduce a class of dissections so that we can state the most important special case of the theorem.

Two dissections are *dual equivalent* if they have the same dual graph. In most applications we are interested in finding an appropriate member of the dual equivalence class. If two dissections are weakly equivalent they need not be dual equivalent, for example in Fig. 4 the rectangle in the lower left corner has 4 neighbors in the left dissection but only 3 in the right dissection.

A segment s of a dissection is *one-sided* if s is the side of at least one of the rectangles, in other words all the segments that have an endpoint on s are on the same side of s. A dissection is *one-sided* if every segment of the dissection is one-sided. The following observation was made in [6].

Proposition 4 *All dissections in the weak equivalence class of a one-sided dissection are dual equivalent.*

Together with Theorem 3 this yields the key theorem.

Theorem 5 *One-sided dissections are area-universal.*

With the following definitions we set the stage for a generalization of Theorem 3. Let $\mu : [0, 1]^2 \to \mathbb{R}_+$ be a density function on the unit square with mass 1, i.e.,

$\int_0^1\int_0^1 \mu(x,y)dxdy = 1$. We assume that μ is well behaved so that all the integrals we need exist and are positive. The *mass* of an axis aligned rectangle $r \subseteq [0, 1]^2$ is defined as $m(r) = \iint_r \mu(x, y)dxdy$.

Theorem 6 *Let $\mu : [0, 1]^2 \to \mathbb{R}_+$ be a density function on the unit square. If R is a dissection with rectangles $r_1, \ldots, r_{n+1}$ and $w : \{1, \ldots, n+1\} \to \mathbb{R}_+$ a positive weight function with $\sum_{i=1}^{n+1} w(i) = 1$, then there exists a unique dissection R' in the unit square that is weakly equivalent to R such that the mass of the rectangle r_i' in R' is exactly $w(i)$.*

A full proof of the theorem is given in [9]. Here we only sketch the proof, it is based on a force directed method that exploits a physical analogy with air-pressure. Consider a representation of R in the unit square and compare the mass $m(r_i)$ to the intended mass $w(i)$. The quotient of these two values can be interpreted as the pressure inside the rectangle. Integrating this pressure along a side of the rectangle yields the force by which r_i is pushing against the segment that contains the side. The difference of pushing forces from both sides of a segment yields the effective force acting on the segment. The intuition is that shifting a segment in the direction of the effective force yields a better balance of pressure in the rectangles. We show that iterating such improvement steps drives the realization of R towards a situation with $m(r_i) = w(i)$ for all i, i.e., the procedure converges towards the dissection R' whose existence we want to show.

Let $r_i = [x_l, x_r] \times [y_b, y_t]$ be a rectangle of R. The pressure $p(i)$ in r_i is $p(i) = \frac{w(i)}{m(r_i)}$. Let s be a segment of R and let r_i be one of the rectangles with a side in s. Let s be vertical with x-coordinate x_s and let $s \cap r_i$ span the interval $[y_b(i), y_t(i)]$. The (undirected) *force* $f(s, i)$ imposed on s by r_i is the pressure $p(i)$ of r_i times the density dependent length of the intersection, i.e.,

$$f(s, i) = p(i) \int_{y_b(i)}^{y_t(i)} \mu_{x_s}(y)dy.$$

The *force* $f(s)$ *acting on s* is obtained as a sum of the directed forces imposed on s by incident rectangles.

$$f(s) = \sum_{r_i \text{ left of } s} f(s, i) - \sum_{r_i \text{ right of } s} f(s, i).$$

Symmetric definitions apply to horizontal segments.

Balance for Rectangles and Segments

A segment s is in *balance* if $f(s) = 0$. A rectangle r_i is in *balance* if $m(r_i) = w(i)$, i.e., if $p(i) = 1$.

Lemma 7 *All rectangles r_i of R are in balance if and only if all segments are in balance.*

Proof We only show one direction. Since all rectangles are in balance we can eliminate the pressures from the definition of the $f(s, i)$. With this simplification we get for a vertical segment s

$$f(s) = \sum_{r_i \text{ left of } s} \int_{y_b(i)}^{y_t(i)} \mu_{x_s}(y)dy - \sum_{r_j \text{ right of } s} \int_{y_b(j)}^{y_t(j)} \mu_{x_s}(y)dy.$$

Hence $f(s) = M_s - M_s = 0$, where M_s is the integral of the fiber density μ_{x_s} along s.

□

Balancing Segments and Optimizing the Entropy

Proposition 8 *If a segment s of R is unbalanced, then we can keep all the other segments at their position and shift s parallel to a position where it is in balance. The resulting dissection R' is weakly equivalent to R.*

The *entropy* of a rectangle r_i of R is defined as $-w(i) \log p(i)$. The *entropy* of the dissection R is

$$E = \sum_i -w(i) \log p(i)$$

The proof of Theorem 6 is in five steps:

(1) The entropy E is always nonpositive.
(2) $E = 0$ if and only if all rectangles r_i of R are in balance.
(3) Shifting an unbalanced segment s into its balance position increases the entropy.
(4) The process of repeatedly shifting unbalanced segments into their balance position makes R converge to a dissection R' such that the entropy of R' is zero.
(5) The solution is unique.

3 Solutions to the Problems

Mapping Segments on Points

Let R be a generic dissection with n segments. Let P be a generic set of n points in a rectangle F. Recall that a *cover map* from R to P is a dissection R' with outer rectangle F that is weakly equivalent to R such that every segment of R' contains exactly one point from P. The following theorem from [9] answers our first problem.

Fig. 6 Dissections R and the dissection R_S obtained by doubling the segments

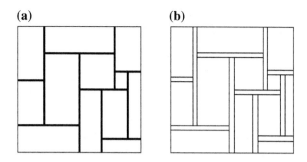

Theorem 9 *If R is a generic dissection with n segments and P is a generic set of n points in a rectangle F, then there is a cover map from R to P.*

To be able to use Theorem 6 we first transform the point set P into a suitable density distribution $\mu = \mu_P$ inside F. This density is defined as the sum of a uniform distribution μ_1 with $\mu_1(q) = 1/\text{area}(F)$ for all $q \in F$ and a distribution μ_2 that represents the points of P. Choose some $\Delta > 0$ such that for all $p, p' \in P$ we have $|x_p - x_{p'}| > 3\Delta$ and $|y_p - y_{p'}| > 3\Delta$, this is possible because P is generic. Define $\mu_2 = \sum_{p \in P} \mu_p$ where $\mu_p(q)$ takes the value $(\Delta^2\pi)^{-1}$ on the disk $D_\Delta(p)$ of radius Δ around p and value 0 for q outside of this disk.

For a density ν over F and a rectangle $r \subseteq F$ we let $\nu(r)$ be the integral of the density ν over r. Using this notation we can write $\mu_1(F) = 1$ and $\mu_p(F) = 1$ for all $p \in P$, hence the total mass of F is $\mu(F) = 1 + n$.

Next we transform the dissection R into a dissection R_S. To this end we replace every segment by a thin rectangle, see Fig. 6. Let $\mathscr{S}$ be the set of new rectangles.

Define weights for the rectangles of R_S as follows. If R_S has r rectangles we define $w(r) = 1 + 1/r$ if $r \in \mathscr{S}$ and $w(r) = 1/r$ for all the rectangles of R_S that came from rectangles of R. Note that the total weight, $\sum_r w(r) = 1 + n$, is in correspondence to the total mass $\mu(R)$.

The data R with μ and R_S with w constitute, up to scaling of R and w, a set of inputs for Theorem 6. From the conclusion of the theorem we obtain a dissection R'_S weakly equivalent to R_S such that $m(r) = \iint_r \mu(x,y) dx dy = w(r)$ for all rectangles r of R'_S.

The definition of the weight function w and the density μ is so that R'_S should be close to a cover map from R to P: Only the rectangles $r \in \mathscr{S}$ that have been constructed by inflating segments may contain a disk $D_\Delta(p)$ and each of these rectangles may contain at most one of the disks. This suggests a correspondence $\mathscr{S} \leftrightarrow P$. However, a rectangle $r \in \mathscr{S}$ can use parts of several discs to accumulate mass. To find a correspondence between $\mathscr{S}$ and P we define a bipartite graph G whose vertices are the points in P and the rectangles in $\mathscr{S}$:

- A pair (p, r) is an edge of G iff $r \cap D_\Delta(p) \neq \emptyset$ in R'_S.

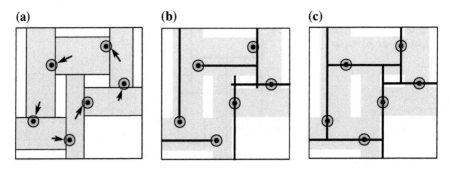

Fig. 7 (a) A solution R'_S with a matching indicated by the arrows. (b) Segments are shifted to the corresponding points. (c) Small final adjustments (clipping and enlarging) yield R'.

The proof of the theorem is completed by proving two claims:

- G admits a perfect matching.
- From R'_S and a perfect matching M in G we can produce a dissection R' that realizes the cover map from R to P.

For the first of the claims we check Hall's matching condition. Consider a subset A of $\mathscr{S}$. Since R_S realizes the prescribed weights we have $m(A) = \mu(A) = \sum_{r \in A} \mu(r) = \sum_{r \in A} w(r) = |A|(1 + 1/r)$. Since $\mu_1(A) < 1$ and $\mu_p(A) \leq 1$ for all $p \in P$, there must be at least $|A|$ points $q \in P$ with $\mu_q(A) > 0$, these are the points that have an edge to a rectangle from A in G. We have thus shown that every $A \subset \mathscr{S}$ is incident to at least $|A|$ points in G, hence, there is an injective mapping $\alpha : \mathscr{S} \to P$ such that $r \cap D_\Delta(\alpha(r)) \neq \emptyset$ in R'_S for all $r \in \mathscr{S}$.

Given the matching α the construction of the dissection R' that realizes the cover map from R to P is completed in two further steps, see Fig. 7b, c.

Cartograms with Optimal Complexity

In a series of papers the complexity of polygons used for the cartograms of triangulations has been reduced from 40 to 34 then 12 and 10. Finally, in [3] the following optimal result was obtained.

Theorem 10 *Every planar triangulation admits an area-universal rectilinear cartogram with ≤ 8-sided polygons.*

The construction is fairly easy with the right tools at hand. First we need a Schnyder wood of the input triangulation G. Let a_1, a_2, a_3 be the outer vertices of G, an orientation and coloring of the inner edges with 3 colors (we identify colors $(1, 2, 3)$ with (red, green, blue)) is a *Schnyder wood* if:

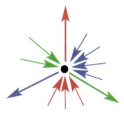

Fig. 8 Schnyder's edge coloring rule

(1) All edges incident to an outer vertex a_i are in-edges and colored i.
(2) Every inner vertex v has three outgoing edges colored red, green and blue in clockwise order. All the incoming edges in an interval between two outgoing edges are colored with the third color, see Fig. 8.

These structures were defined by Schnyder in [15], where it was also shown that every triangulation admits a Schnyder wood. Moreover, if T_i is the set of oriented edges of color i and T_i^{-1} is the same set with reversed orientations, then it holds that $T_1 \cup T_2^{-1} \cup T_3^{-1}$ is acyclic. This property can be used to show that every triangulation has a contact representation with internally disjoint $\perp$ shapes. Figure 9 shows an example.

The $\perp$-representation can be viewed as a rectangular dissection. Now replace every segment of this dissection by a thin rectangle. This yields a one-sided dissection R_G, see Fig. 10(left). With a vertex v of G we associate the polygon P_v formed as the union of four rectangles. These are the two rectangles that were obtained from the two segments of the $\perp$ shape representing v together with the two rectangles that have parts of the horizontal segment of this $\perp$ as bottom side. In Fig. 10(right) the polygons P_v are distinguished by colors.

It is easily checked that the polygons P_v have at most 8 corners, hence, at most 8 sides. Given a set of weights $w : V \to \mathbb{R}^+$ we can arbitrarily break $w(v)$ into four positive values and assign these to the rectangles constituting P_v. Since the dissection R_G is one-sided and, hence, area-universal there is a realization of the dissection where the area of $P(v)$ equals $w(v)$.

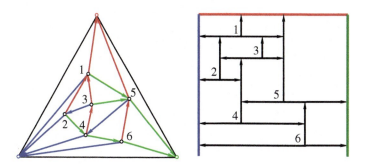

Fig. 9 A triangulation with a Schnyder wood and a $\perp$-representation

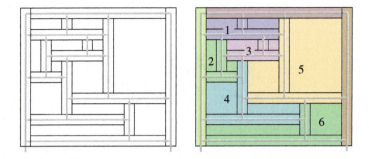

Fig. 10 The one-sided dissection resulting from the ⊥-representation of Fig. 9

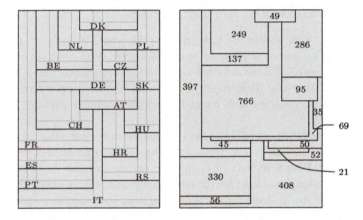

Fig. 11 Central European states represented by polygons with at most 8 sides (*left*). A cartogram where the areas are proportional to the emission of CO_2 in 2009 (*right*).

From the thesis of Torsten Ueckerdt [20] we borrow Fig. 11, which shows a cartogram displaying real data. The cartogram was computed with the method of this section.

4 Background and Additional Problems

Theorem 3 was first proven by Wimer et al. [21]. They take the width x_i and height y_i of the rectangle r_i as variables and show that the system consisting of linear equations which correspond to left-to-right and bottom-to-top sequences of rectangles together with the non-linear equations $x_i y_i = w(i)$ has a unique solution. The theorem was rediscovered by Eppstein et al. [5]. They prove it with an argument based on *"invariance of domain."* Both proofs are purely existential. However, in [5, 6] it is noted that the solution can be computed by iteratively reducing the distance between the present weights and the intended weight vector $(w(i))_i$. The 'air-pressure' method proposed by Izumi, Takahashi and Kajitani [10] is such an iterative approach. In [9] the air-pressure technique was used to prove Theorem 6. The sketch given here is based on this paper. A short non-constructive proof of Theorem 6 was given by Schrenzenmaier [16, p. 21], he adapted the proof of Theorem 3 from [5].

The two main problems regarding area-universal rectangular dissections are the following:

- Given R and w, is it possible to compute the weakly equivalent dissection R' realizing the weights in polynomial time (efficient Theorem 3)?
- Characterize graphs that admit a one-sided dual or find a polynomial recognition algorithm for them.

Beside this it would be very interesting to identify further instances of area-universality. Two such instances are straight line drawings of 3-regular planar graphs with prescribed face areas (Thomassen [18]) and straight line drawings of grids with prescribed face areas, a.k.a. table cartograms (Evan et al. [7]).

Problem 1 was a conjecture of Ackerman et al. [1]. They were motivated by the study of the function $Z(P)$ counting rectangulations of a generic point set P. Combining results from [9] (lower) and [1] (upper) we know that $Z(P)$ is in $\Omega(8^{n+1}/(n+1)^4)$ and in $O(20^n/n^4)$. The lower bound is tight for some sets P, to improve the upper bound remains an intriguing problem.

The construction of area-universal rectilinear cartograms with ≤ 8-sided polygons is from Alam et al. [3]. As already noted the construction based on our key theorem is not known to be efficient. Polynomial constructions of cartograms with ≤ 8-sided polygons are known for Hamiltonian triangulations [3] and with ≤ 10-sided polygons for general triangulations [2]. Is there a polynomial algorithm for constructing cartograms with ≤ 8-sided polygons for general triangulations?

Recognition of planar graphs which admit rectangular cartograms or cartograms with ≤ 6-sided polygons is also wide open.

References

1. Ackerman, E., Barequet, G., Pinter, R.Y.: On the number of rectangulations of a planar point set. J. Comb. Theory Ser. A **113**(6), 1072–1091 (2006)
2. Alam, M.J., Biedl, T., Felsner, S., Gerasch, A., Kaufmann, M., Kobourov, S.G.: Linear-time algorithms for rectilinear hole-free proportional contact representations. Algorithmica **67**, 3–22 (2013)
3. Alam, M.J., Biedl, T., Felsner, S., Kaufmann, M., Kobourov, S.G., Ueckerdt, T.: Computing cartograms with optimal complexity. Discrec. Comput. Geom. **50**, 784–810 (2013)
4. Dasgupta, P., Sur-Kolay, S.: Slicible rectangular graphs and their optimal floorplans. ACM Trans. Des. Autom. Electron. Syst. **6**(4), 447–470 (2001)
5. Eppstein, D., Mumford, E., Speckmann, B., Verbeek, K.: Area-universal rectangular layouts, 19 pp. (2009). arXiv:0901.3924
6. Eppstein, D., Mumford, E., Speckmann, B., Verbeek, K.: Area-universal and constrained rectangular layouts. SIAM J. Comput. **41**, 537–564 (2012)
7. Evans, W., Felsner, S., Kaufmann, M., Kobourov, S.G., Mondal, D., Nishat, R.I., Verbeek, K.: Table cartograms. In: Proceedings ESA. Lecture Notes in Computer Science, vol. 8242, pp. 421–432. Springer (2013)
8. Felsner, S.: Rectangle and square representations of planar graphs. In: Pach, J. (ed.) Thirty Essays in Geometric Graph Theory, pp. 213–248. Springer, New York (2013)
9. Felsner, S.: Exploiting air-pressure to map floorplans on point sets. J. Graph Algorithm Appl. **18**, 233–252 (2014)
10. Izumi, T., Takahashi, A., Kajitani, Y.: Air-pressure model and fast algorithms for zero-wasted-area layout of general floorplan. IEICE Trans. Fundam. Electron., Commun. Comput. Sci. **E81-A**, 857–865 (1998)
11. Nassar, K.: New advances in the automated architectural space plan layout problem. In: Proceedings Computing in Civil and Building Engineering, 9 pp. (2010). http://www.engineering.nottingham.ac.uk/icccbe/proceedings/pdf/pf193.pdf
12. Nishizeki, T., Rahman, M.S.: Planar Graph Drawing. Lecture Notes Series on Computing. World Scientific, Hackensack (2004)
13. Raisz, E.: The rectangular statistical cartogram. Geogr. Rev. **24**(3), 292–296 (1934)
14. Rinsma, I.: Non-existence of a certain rectangular floorplan with specified area and adjacency. Environ. Plan. **14**, 163–166 (1987)
15. Schnyder, W.: Planar graphs and poset dimension. Order **5**, 323–343 (1989)
16. Schrenzenmaier, H.: Ein Luftdruckparadigma zur Optimierung von Zerlegungen. Bachelor's thesis, TU Berlin (2013)
17. Team, W.: Worldmapper, The world as you've never seen it before. http://www.worldmapper.org
18. Thomassen, C.: Plane cubic graphs with prescribed face areas. Comb. Prob. Comput. **1**, 371–381 (1992)
19. Tobler, W.: Thirty five years of computer cartograms. Ann. Assoc. Am. Geogr. **94**, 58–73 (2004)
20. Ueckerdt, T.: Geometric representations of graphs with low polygonal complexity. Dissertation, TU Berlin (2011)
21. Wimer, S., Koren, I., Cederbaum, I.: Floorplans, planar graphs, and layouts. IEEE Trans. Circuits Syst. **35**(3), 267–278 (1988)

Linear Structure of Graphs and the Knotting Graph

Ekkehard Köhler

Abstract Many important graph classes, such as interval graphs, comparability graphs and AT-free graphs, show some kind of linear structure. In this paper we try to capture the notion of linearity and show some algorithmic implications. In the first section we discuss the notion of linearity of graphs and give some motivation for its usefulness for particular graph classes. The second section deals with the knotting graph, a combinatorial structure that was defined by Gallai long ago and that has various nice properties with respect to our notion of linearity. Next we define intervals of graphs in Sect. 3. This concept generalizes betweenness in graphs—a crucial notion for capturing linear structure in graphs. In the last section we give a practical example of how to use the linear structure of graphs algorithmically. In particular we show how to use these structural insights for finding maximum independent sets in AT-free graphs in $O(n\overline{m})$ time, where $\overline{m}$ denotes the number of non-edges of the graph G.

1 Linear Structure of Graphs

There are many combinatorial problems that are NP-hard on graphs in general but need to be solved in practice. Often people give up on their search for an optimal solution and are satisfied with solutions that at least approximate the optimum by a not too bad factor. For various applications this might be indeed sufficient but it still is rather unsatisfying to be left with a solution that can be proven to be "only" a constant factor away from the optimum. Yet, there is an alternative approach by not relaxing optimality but instead more carefully studying the structure of the input. In many cases the input graph is not of arbitrary structure but rather has some very helpful properties that can be used algorithmically to find optimal algorithms for this particular input although the problem in general is still NP-hard. In this paper we study such a structural property that has been helpful in many cases. We investigate

E. Köhler (✉)
Mathematisches Institut, Brandenburgische Technische Universität,
Platz der Deutschen Einheit 1, 03046 Cottbus, Germany
e-mail: ekkehard.koehler@b-tu.de

Fig. 1 Three graphs that show some kind of linear structure

how to find out whether the given input has this particular property and we will show how to make use of this property algorithmically.

Graphs with Linear Structure

Consider the three examples in Fig. 1. The first graph is an interval graph, i.e., there is a set of (closed) intervals on the real line, such that for each vertex of the graph there is exactly one such interval and two vertices are adjacent in the interval graph if the corresponding intervals have a non-empty intersection. The second graph shows a poset (a partially ordered set), where the vertices are the elements of this poset and two elements are comparable when there is an (oriented) edge between them. The third example shows an AT-free graph. This means that the graph does not contain a particular kind of structure that is called asteroidal triple as an induced subgraph. An asteroidal triple (AT) in a graph G is a set of three vertices v_1, v_2, v_3 such that for each $1 \leq i \leq 3$ the vertices $\{v_1, v_2, v_3\} \setminus \{v_i\}$ are contained in the same connected component of $G \setminus N[v_i]$. Here $N[v]$ denotes the closed neighborhood of v in G, i.e., the set of neighbors of vertex v including v itself; $N(v)$ will be used for the set of neighbors of v excluding v. See Fig. 2 for three small graphs having an asteroidal triple.

The three example graphs in Fig. 1 appear to have a completely different structure. However, a closer look shows that they all share the property that there is some kind of underlying linear structure in these graphs. For the interval graph the linearity is inherited by the simple structure of the real line; for the partial order the linearity is induced by the transitivity of the partial order relation; and for the AT-free graph this linear structure is not that easy to recognize yet, but, as will be explained more detailed in the course of this paper, the non-existence of ATs restricts the graph in some sense to extend only in two different directions. These very different interpretations of linearity seem to be unrelated at first sight. But a closer look reveals that there are indeed some underlying strong structural relationships between these concepts.

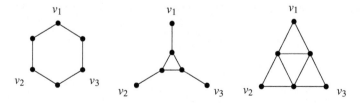

Fig. 2 Three small graphs, each having an asteroidal triple on the vertices v_1, v_2, v_3

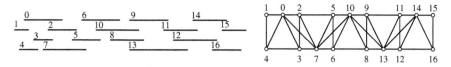

Fig. 3 An interval model together with the corresponding interval graph

Algorithmic Implications of Linearity

Suppose that we are indeed able to show some linearity of our given graph. What is it good for? Can such a linear structure be used algorithmically? Indeed, for various problems this can be shown to be the case. To get an idea how linear structure can be utilized to solve combinatorial problems in a very simple way, consider our first class of examples, the interval graphs and the optimization problem to find a maximum weighted independent set. Let the interval graph be given together with an interval model (see Fig. 3 for an example). Now apply the following simple algorithm:

- take the interval model sorted by increasing right endpoint and scan the interval model from left to right
- when the first point of some interval i is reached: store the weight of i plus the weight of the largest interval that has been closed before as the new weight of i
- when the last point of some interval i is reached: put its weight into the (ordered) list of closed intervals

Given the interval model, it is easy to implement this algorithm in linear time. Also, to show that the algorithm finds the optimal solution is not hard: Assume there is a larger independent set and then looking at the first point where the two independent sets differ in the scanning procedure described above. Obviously, the linear structure has been utilized by scanning the geometric model from left to right. But there is also another way to interpret the linear structure here. Consider the complement graph of our interval graph G. This complement graph is the underlying graph of the partial order on the intervals, where an interval i is less than or equal to j, if i's interval is left of the interval of j. In this partial order the scanning algorithm can be interpreted as iteratively visiting the minimal elements of the partial order and, simultaneously updating their weight function. Graphs that are underlying undirected graphs of partial orders are called comparability graphs and their complements are so-called cocomparability graphs. Building up on the above algorithmic idea, one can design a linear time algorithm for the maximum weighted clique problem for comparability graphs [10]. In fact, as recently shown [9], a similar idea can be used to design a linear time algorithm for the maximum weight independent set problem on cocomparability graphs, i.e., a super-class of interval graphs. It is well-known and not difficult to see (see Theorem 5) that the class of AT-free graphs is a super-class of cocomparability graphs. A natural question is whether it is possible to apply similar algorithmic approaches based on the linear structure also to AT-free graphs. Unfortunately, this is not possible in such an easy way. Still also here the linear structure helps to find algorithms. Broersma et al. [1] were the first to show that a polynomial time algorithm for the maximum weight independent set problem exists, with running time $O(n^4)$. We will show later how to use some insight on the linear structure to improve their algorithm to get a running time of $O(n^3)$.

Notions of Linearity

Before we can concentrate on those algorithmic questions we should first get a better idea of what the linear structure of a graph is supposed to be. In particular, why is it appropriate to consider AT-free graphs to have such a structure. There are various properties that suggest some kind of linear property of AT-free graphs. A first one is due to Corneil et al. [3]. They showed that in every AT-free graph there is a dominating pair, i.e., there is a pair of vertices x, y such that every induced path between these two vertices dominates the whole graph. In that sense every such path can be seen as a certificate for the linear structure of the graph and the vertices x and y as some kind of "end-vertices" of the graph. Another way to argue for some kind of linear structure of AT-free graphs is a result by Möhring [11]. He showed that every minimal triangulation of an AT-free graph is an interval graph. More precisely, with a beautiful simple method he showed that for an AT-free graph G and an inclusion minimal set of edges such that G plus these edges does not have an induced cycle without a chord, the resulting graph is always an interval graph. Thus, no matter how one destroys induced cycles in an AT-free graph by inserting chords, the resulting graph has always an obvious linear property that can be seen in the corresponding simple geometric interval model. Later Parra and Scheffler [12] and, independently Corneil et al. [3], proved that Möhring's result even gives a characterization of AT-free graphs.

Up to here we have not been very clear about what we mean by linear structure in a graph and, indeed, this is rather hard to do in a very general setting. Therefore we try to get a bit more formal in the following. One property of any definition of linearity is some kind of ordering of the vertices of a given graph. In particular one should be able to determine at least for some pairs of vertices, which of the two vertices appears before the other vertex in the linear ordering. The class of graphs where this ordering property is most evident is the class of comparability graphs. Here this property implies a lot of structural properties for the graph class. In fact, as we will see next, there is a very helpful combinatorial structure, the knotting graph, that is capable to captures the properties of the ordering in a very elegant way.

2 Knotting Graphs and Their Properties

Let P be a partially ordered set (poset) on a set V, i.e., P can be interpreted as a subset of $V \times V$, representing a transitive, antisymmetric and irreflexive relation $\leq$, and if (a, b) is one of these ordered pairs then this means that $a \leq b$ in P. Obviously this poset can be interpreted as an oriented graph; the underlying undirected graph is called a comparability graph, or, alternatively, a transitively orientable graph. Reversely, assume now to be given an undirected graph G. The aim is to find out whether there is a partial order P such that G is the underlying undirected graph of

P. There are various methods for solving this recognition problem efficiently. One of the most elegant approaches follows from a result of Gallai [5] and is based on the following idea. Consider some edge uv of the graph G under consideration. If there is a transitive orientation of G, then uv has to be oriented from u to v or from v to u. Without loss of generality, we can assume uv to be oriented from u to v. If there is another edge uw in G such that $vw \notin E$ then, by transitivity, also uw has to be oriented from u to w. In that sense one can say that the orientation of uv *forces* the orientation of uw. More generally, let us only look at the edges incident to vertex u. We can conclude by the same argument as above that the orientation of uv forces the orientation of all edges $uw \in E$ if v and w are in the same connected component of $\overline{G}[N(u)]$. Here $\overline{G}[N(u)]$ denotes the graph induced by the neighborhood of v in the complement of G. Thus there is a path of non-edges between v and w in $N(v)$. One could say those edges are *knotted* together at this vertex u. Building on this idea Gallai suggested a graph structure, called the knotting graph that is capable to capture this local forcing property.

Definition 1 For a given graph $G = (V, E)$ the corresponding *knotting graph* is given by $K[G] = (V_K, E_K)$ where V_K and E_K are defined as follows. For each vertex v of G there are copies $v_1, v_2, \ldots, v_{i_v}$ in V_K, where i_v is the number of connected components of $\overline{N(v)}$, the complement of the graph induced by $N(v)$. For each edge (v, w) of E there is an edge (v_i, w_j) in E_K, where w is contained in the ith connected component of $\overline{N(v)}$ and v is contained in the jth connected component of $\overline{N(w)}$.

Example 2 In Fig. 4 one can see a graph G together with its knotting graph. Here small dots in the knotting graph that are drawn closely together indicate that they are copies of the same original vertex of the graph.

Obviously, the number of edges of $K[G]$ is the same as the number of edges of G, whereas the number of vertices of $K[G]$ is in the order of the number of edges of G.

Although the above mentioned forcing relation is defined purely locally for each vertex separately, the knotting graph allows to look at this forcing on a more global scale. Two edges are said to *force each other* if there is a sequence of forcings

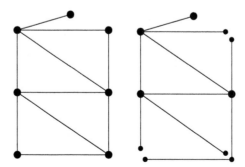

Fig. 4 Example of a graph, together with its knotting graph

between those edges, as explained above. That way the edges of the whole graph can be partitioned into edge-classes such that two edges are in the same class if they force each other. Gallai observed that the knotting graph of a given graph G basically represents the edge classes of G in the sense that the connected components of the knotting graph are the edge classes of the original graph. Already Gilmore and Hoffmann indicated the relevance of these edge-classes when they stated their famous theorem to characterize comparability graphs [6]. Much later, Kelly [7] also studied the knotting graph and, among others, suggested a simple algorithm to construct the modular decomposition of a graph, using its knotting graph. In short: The knotting graph turned out to be quite useful.

For the purpose of characterizing comparability graphs, i.e., the class of graphs which have a transitive orientation, the knotting graph has special importance, as shown in the following theorem.

Theorem 3 [5] *A graph G is transitively orientable if and only if* $K[G]$ *is bipartite.*

To see the one direction of this theorem it is sufficient to observe that an induced odd cycle with more than three vertices does not have a transitive orientation and thus any odd cycle in the knotting graph is an obstruction to a transitive orientation. The reverse direction is a bit more involved; here Gallai uses a powerful structural result on modular decompositions of comparability graphs. For this direction we refer the reader to the original paper by Gallai [5].

Having seen that the knotting graph captures transitive orientations and thus the linear structure of comparability graphs in a very natural way, it seems plausible to ask whether for related, more general classes of graphs similar properties can be shown. As mentioned earlier the class of graphs that seems to have some linear properties is the class of AT-free graphs. When studying this class more carefully then it becomes clear that we in fact should study their complement, the so-called coAT-free graphs since they are the generalization of comparability graphs, whereas AT-free graphs generalize cocomparability graphs. Already Gallai proved a strong relationship between these two classes by giving a characterization of comparability graphs using a list of forbidden subgraphs. Some of these graphs are asteroidal triples in the complement. The question is of course, whether there is a similarly simple characterization of coAT-free graphs using the knotting graph, as there is for comparability graphs. And indeed such a result exists. Even more, one can generalize the concepts of asteroidal triple to asteroidal sets and still get such a characterization. Before we can state this result we first have to define the concept of an asteroidal set and the asteroidal number of a graph G.

Definition 4 For a given graph G, an independent set of vertices S is called *asteroidal set* if for each $x \in S$ the set $S - \{x\}$ is in one connected component of the graph $G - N[x]$. The *asteroidal number* of a graph G is defined as the maximum cardinality of an asteroidal set of G, and is denoted by $an(G)$.

The theorem itself is now rather simple.

Theorem 5 [8] *Let G be a graph, then* $\mathrm{an}(G) = \omega(\mathrm{K}[\overline{G}])$.

Proof Let $\mathrm{an}(G) = k$ and let $A = \{a_1, \ldots, a_k\}$ be an asteroidal set of G. By the definition of asteroidal sets the vertices of A are pairwise independent. Consequently, A induces a clique in $\overline{G}$ and for each $a_i \in A$ the set $A \setminus \{a_i\}$ is contained in the neighborhood of a_i in $\overline{G}$. Since A is an asteroidal set, for each $a_j, a_k \in A \setminus \{a_i\}$ ($j \neq k$) there is an a_j, a_k-path in G that avoids the neighborhood of a_i. Therefore a_j and a_k are in the same connected component of $G - N[a_i]$. By the knotting graph definition this implies that the knotting graph edges corresponding to the edges $a_i a_j, a_i a_k$ of $\overline{G}$ are incident to the same copy of a_i in the knotting graph. Since this is true for all pairs of vertices in $A \setminus \{a_i\}$, all edges corresponding to edges from vertices of $A \setminus \{a_i\}$ to a_i in $\overline{G}$, are incident to the same copy of a_i in the knotting graph. Consequently, there is a k-clique in $\mathrm{K}[\overline{G}]$ formed by copies of vertices of A.

Now suppose there is a k-clique in the knotting graph $\mathrm{K}[\overline{G}]$. Since there is a 1-1 correspondence between the edges of $\overline{G}$ and the edges of $\mathrm{K}[\overline{G}]$ there is a set $A = \{a_1, \ldots, a_k\}$ of k vertices of G corresponding to the vertices of the clique in $\mathrm{K}[\overline{G}]$. By the definition of the knotting graph, for each vertex $a_i \in A$ the vertices of $A \setminus \{a_i\}$ are contained in the same connected component of $G - N[a_i]$. Consequently, A is an asteroidal set of G. □

As a simple corollary of this theorem we can conclude that a graph is coAT-free if its knotting graph is triangle-free. Since every bipartite graph is, of course, triangle-free, we also have a very simple proof that every comparability graph is a coAT-free graph.

The initial intention to study some kind of linear structure was the hope to be able to use this structure for solving algorithmic questions. So the question here would be whether the knotting graph is of any help in this regard. Indeed various properties of a graph can be determined much easier when using the knotting graph. As a warm-up consider the problem of finding a dominating pair in a graph. These special pairs of vertices can be found using the following simple proposition.

Proposition 6 *The pair of vertices a, b is a dominating pair in $\overline{G}$ if and only if each common neighbor x of a and b in G has different vertex copies in $\mathrm{K}[G]$ adjacent to the copy of a and the copy of b (i.e., xa and xb are not knotted at x).*

We leave the proof of this proposition to the reader as a simple exercise to get more familiar with the knotting graph.

3 Intervals in Graphs

One crucial property of linear structures is the notion of *betweenness*. If for a vertex v there is some vertex u to the left of v and some vertex w to the right of v in some kind of linear model then one could say that v is *between* u and w, or in other words,

v is in the *interval* between u and w. Broersma et al. [1] defined the notion of an interval with this meaning.

Definition 7 Let $G = (V, E)$ be a connected graph. A vertex $s \in V \setminus \{x, y\}$ is said to be *between* x and y if x and s are in the same component of $G - N[y]$ and y and s are in the same component of $G - N[x]$. The *interval* $I(x, y)$ of G is defined to be the set of all vertices of G that are between x and y.

Let $C^x(y)$ be the component of $G - N[x]$ that contains y. Observe that this definition implies $I(x, y) = C^x(y) \cap C^y(x)$.

Let $K[\overline{G}]$ be the knotting graph of $\overline{G}$. From the definition of the knotting graph it follows directly that for a vertex x of G, for each component of $G - N[x]$ there is a copy of x in $K[\overline{G}]$. Furthermore, for two vertices u, v of G that are adjacent to x in $\overline{G}$, and that are contained in the same connected component C of $G - N[x]$, the corresponding edges in the knotting graph are adjacent to the same copy of x. Hence, the interval $I(x, y)$ corresponds to the set of all vertices u such that the edge from u to x is incident to the same copy of x as the edge from y to x, and the edge from u to y is incident to the same copy of y as the edge from x to y.

Before we study intervals of AT-free graphs more carefully, let us first observe that these intervals have nice properties for comparability graphs as well.

Proposition 8 Let G be a comparability graph with $x, y \in V$, $(x, y) \in E$ and let $I(x, y)$ be the interval of x and y in $\overline{G}$. Then, in any transitive orientation F of G, the vertices of $I(x, y)$ are between x and y, i.e., for each $z \in I(x, y)$ either x is a predecessor of z and z is a predecessor of y, or y is a predecessor of z and z is a predecessor of x (i.e., $x \leq z \leq y$ or $y \leq z \leq x$).

Proof Let G be a comparability graph with a transitive orientation F and, without loss of generality, let x be a predecessor of y. Suppose there is some vertex $z \in I(x, y)$ which is a predecessor of x. By the definition of $I(x, y)$ there has to be a path $P = v_1, v_2, \ldots, v_k$ in $\overline{G} - N_{\overline{G}}[x]$, with $v_1 = z$, $v_k = y$ and $k \geq 3$. Obviously, none of the vertices of P is adjacent to x in $\overline{G}$. Since vertex v_2 is a neighbor of z on P, it is not adjacent to z in G. Thus, v_2 has to be a predecessor of x in F and by transitivity, cannot be adjacent to y in $\overline{G}$ implying $k > 3$. By similar arguments one can show that v_{k-1} has to be a successor of x in F. By the transitivity of F, v_2 is not a neighbor of v_{k-1} in P.

Similarly we can go on in constructing vertices of P. All of them have to be adjacent to vertex x in G and, since F is a transitive orientation of G, each of those vertices either has to be a predecessor or a successor of x. Consequently, the set of vertices of P can be partitioned into two non-empty sets S_P and S_S, the predecessors and the successors of x. Because of the transitivity of F, each vertex of S_P is adjacent to each vertex of S_S. Hence those vertices cannot form a connected path in $\overline{G}$. □

Let us return to AT-free graphs. Broersma et al. showed some properties of intervals in AT-free graphs. We can use the knotting graph to prove and extend these properties in a straight-forward way.

Proposition 9 [1] *Let s be an element of $I(x, y)$. Then x and y are in different components of $G - N[s]$.*

Proposition 10 [1] *For $s \in I(x, y)$ we have $I(x, s) \cap I(s, y) = \emptyset$.*

If we use the knotting graph these two properties can be seen easily. The first one follows directly from Theorem 5, as the knotting graph of an asteroidal triple-free graph is triangle-free. Proposition 10 follows from the fact that for each edge of $\overline{G}$ there is only one copy of this edge in the knotting graph.

By Proposition 9, for each vertex s of $I(x, y)$ the edges of $\overline{G}$, connecting s to vertex x and vertex y, are adjacent to different copies of s in the knotting graph. That is, vertex s *separates* x and y in G in the sense that in $G - N[s]$ vertex x and vertex y are in different connected components.

Consequently, any vertex r that has in $K[\overline{G}]$ an edge to the same copy of s as vertex x, is adjacent to the same copy of y as is x. With this observation the next lemma follows immediately.

Lemma 11 [1] *For $s \in I(x, y)$ we have $I(x, s) \subset I(x, y)$ and $I(s, y) \subset I(x, y)$.*

Using the interpretation of intervals in the knotting graph we can prove the following characterization of AT-free graphs.

Theorem 12 *A graph G is AT-free if and only if for each interval $I(x, y)$ of G and each $z \in I(x, y)$ we have $I(x, z) \subseteq I(x, y)$ and $I(z, y) \subseteq I(x, y)$.*

Unfortunately, the meaning of "between" for AT-free graphs is not as clear as for comparability or cocomparability graphs. As the next example shows, the edges of odd holes in the knotting graph of an AT-free graph can all correspond to non-empty intervals.

Example 13 Let G be the graph drawn as the left graph in Fig. 5. The red vertices are assumed to form a complete graph minus the edges of an induced cycle, containing all red vertices (the edges of the cycle are indicated by the dashed lines). The blue vertices are assumed to form a complete graph. The graph on the right-hand side of Fig. 5 represents the knotting graph of $\overline{G}$. Since $K[\overline{G}]$ is triangle-free, G is an AT-free graph. As one can check easily, each edge of the induced odd cycle in $K[\overline{G}]$ represents a non-empty interval of G, implying that there is an odd cycle of intervals that are non-empty and pairwise disjoint.

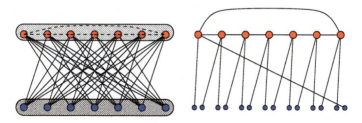

Fig. 5 An AT-free graph G together with $K[\overline{G}]$ containing an odd cycle of non-empty intervals

The following theorem is proved in [1].

Theorem 14 [1] *Let G be an AT-free graph, let $I(x, y)$ be an interval of G, and let s be an element of $I(x, y)$. There exist components $C_1^s, C_2^s, \ldots, C_t^s$ of $G - N[s]$ such that*

$$I(x, y) \setminus N[s] = I(x, s) \cup I(s, y) \cup \bigcup_{i=1}^{t} C_i^s. \tag{1}$$

In the terminology of the knotting graph $K[\overline{G}]$ corresponding to an AT-free graph G, this theorem means the following. Let s' be a copy of s in $K[\overline{G}]$ that is neither incident to the edge from x to s nor to the edge from y to s and let $N(s')$ be the set of neighbors of s' in $K[\overline{G}]$. Then, by Theorem 14 the set $N(s')$ is either completely contained in $I(x, y)$ or it does not contain any element of $I(x, y)$.

Let $\mathscr{C}^s = \{C_i^s : C_i^s \text{ component of } G - N[s]\}$ be the set of all components of $G - N[s]$. The next theorem shows that Theorem 14 can be sharpened in the sense that we can characterize the set of components $C_1^s, \ldots, C_t^s$. In the notation of the knotting graph this theorem says that for each s' of $K[\overline{G}]$ that is neither incident to the edge from x nor to the edge from y, the set $N(s')$ either is contained completely within $I(x, y)$ or there is some copy x' of x such that $N(s') = N(x')$ or some copy y' of y such that $N(s') = N(y')$.

Theorem 15 *Let G be an AT-free graph, let $I(x, y)$ be an interval of G, let s be a vertex of $I(x, y)$, and let $C_1^s, C_2^s, \ldots, C_t^s$ be the components of $\mathscr{C}^s \setminus (\mathscr{C}^x \cup \mathscr{C}^y \cup \{C^s(x), C^s(y)\})$. Then the following holds*

$$I(x, y) \setminus N[s] = I(x, s) \cup I(s, y) \cup \bigcup_{i=1}^{t} C_i^s. \tag{2}$$

Proof Let $C^s(z)$ be the component of $G - N[s]$ that contains z, with $z \notin I(x, y)$ and $z \notin C^s(x) \cup C^s(y)$. By Theorem 14 no vertex p of $C^s(z)$ is contained in $I(x, y)$, since $z \notin I(x, y)$. Furthermore, since $z \notin C^s(x) \cup C^s(y)$, no vertex p of $C^s(z)$ is contained in $N[x] \cup N[y]$.

Vertex z is not contained in $I(x, y)$. Hence, either y and z are in different connected components of $G - N[x]$ or x and z are in different connected components of $G - N[y]$. Without loss of generality, we assume the first case holds. All we have to show now is that $C^s(z)$ is a component of $G - N[x]$.

Since $C^s(z)$ induces a connected graph and none of the vertices of $C^s(z)$ is contained in the neighborhood of x, all vertices of $C^s(z)$ are contained in the same component $C^x(z)$ of $G - N[x]$ and y is not contained in $C^x(z)$.

Suppose there is some vertex q in $C^x(z)$ that is not contained in $C^s(z)$. Among all those q we select one that has a neighbor in $C^s(z)$, which has to exist since $C^x(z)$ is a connected component. Obviously, q is not adjacent to x, since $C^x(z)$ is a component of $G - N[x]$. If q is not adjacent to s either, then $s \in G - N[s]$ and, since q has a neighbor in $C^s(z)$, it holds that $q \in C^s(z)$ which is a contradiction. Consequently, q is adjacent to s. Since, on the other hand, s is not contained in $N[x]$,

$s \in G - N[x]$ and thus $s \in C^x(z)$. By the definition of $I(x, y)$, for all $s \in I(x, y)$ we have $s \in C^x(y)$. Consequently, $C^x(z) = C^x(y)$, contradicting our assumption that z and y are in different components of $G - N[x]$. □

Hence, we have shown that each of the components of $\mathscr{C}^z \setminus \{C^s(x), C^s(y)\}$ that is not contained in $I(x, y)$ is either contained in $\mathscr{C}^x$ or in $\mathscr{C}^y$.

The following simple lemma will be helpful for deriving the algorithm in the following section.

Lemma 16 *Let s be an element of $I(x, y)$ and let C be a component of both $G - N[x]$ and $G - N[y]$. Then C is a component of $G - N[s]$.*

Proof Vertex s does not have a neighbor in C, because otherwise C would be contained in $C^x(y) = C^x(s)$.

Since C is a component of both $G - N[x]$ and $G - N[y]$, all vertices of $N(C)$ have to be contained both in $N[x]$ and $N[y]$. Suppose there is a vertex p in $N(C)$ that is not contained in $N[s]$. Vertex p is adjacent both to x and to y. Consequently there is an x, y-path in $G - N[s]$. But this is a contradiction to Proposition 9. □

4 Improved Algorithm for Independence Number

In this section we show how to apply the linear structure, visible in the knotting graph, to solve the independent set problem in AT-free graphs. In particular we show how to improve the algorithm for computing the independence number of an AT-free graph by Broersma et al. [1]. While their algorithm has a running time of $O(n^4)$, this alteration of the algorithm leads to a running time of $O(n^3)$ or, more precisely $O(n\overline{m})$, where $\overline{m}$ denotes the number of non-edges of the graph G. To explain the improvement, we first state the main idea of the algorithm of [1] and then show how it can be improved. A preliminary version of the improved algorithm was given in the author's PhD thesis.

The basic idea of the algorithm of Broersma is a dynamic programming approach. For a given graph G, the independence number can be expressed by

$$\alpha(G) = 1 + \max_{x \in V} \left(\sum_{i=1}^{r(x)} \alpha(C_i^x) \right), \tag{3}$$

where $C_1^x, C_2^x, \ldots, C_{r(x)}^x$ are the connected components of $G - N[x]$. For a given component C^x of $G - N[y]$, Broersma et al. showed the following equation.

$$\alpha(C^x) = 1 + \max_{y \in C^x} \left(\alpha(I(x, y)) + \sum_i \alpha(D_i^y) \right), \tag{4}$$

where the D_i^y's are the components of $G - N[y]$ contained in C^x. For a given interval $I(x, y)$ of G a similar equation holds. If $I(x, y) = \emptyset$ then $\alpha(I(x, y)) = 0$. Otherwise,

$$\alpha(I(x, y)) = 1 + \max_{s \in I(x,y)} \left(\alpha(I(x, s)) + \alpha(I(s, y)) + \sum_i \alpha(C_i^s) \right), \quad (5)$$

where the C_i^s's are the components of $G - N[s]$ contained in $I(x, y)$. Now, the algorithm of [1] does the following.

Step 1: For every $x \in V$ compute components $C_1^x, \ldots, C_{r(x)}^x$ of $G - N[x]$.
Step 2: For every pair x, y of non-adjacent vertices, determine $I(x, y)$.
Step 3: Sort components and intervals according to their size.
Step 4: Compute $\alpha(C)$ and $\alpha(I)$ for each component and interval in the order of their size, according to the formulas (4) and (5).
Step 5: Compute $\alpha(G)$.

All parts of the algorithm can be shown to run in $O(n^3)$, in fact, a more careful analysis reveals that all those steps take $O(n\overline{m})$ time. There is only one exception—the computation of the independence number of the intervals in Step 4. In [1] this computation is done as follows.

For each vertex s of $I(x, y)$ all vertices $z \in I(x, y) \setminus N[s]$ are considered and if there is a connected component C of $G - N[s]$ that contains z and that was neither already found for some other z', nor the component contains x or y, then the corresponding independence number of C is added to the independence number of $I(x, y)$ for the case that s is an element of the independent set. Hence, for all intervals of G this takes

$$\sum_{\{x,y\} \notin E} \sum_{s \in I(x,y)} O(|I(x, y)|) = O(n^4).$$

Before we start to explain the altered version of the algorithm, we make a simple observation.

Proposition 17 *Let $G = (V, E)$ be a graph and let $\mathscr{C}$ be the set of components of G that, for some vertex $x \in V$, are components of $G - N[x]$. Then $\mathscr{C}$ has no more than $2\overline{m}$ elements.*

Proof Let $K[\overline{G}]$ be the knotting graph, corresponding to G. Obviously, every element of $\mathscr{C}$ corresponds to at least one edge of the knotting graph and, on the other hand, there are no more than two elements of $\mathscr{C}$ that correspond to an edge e of the knotting graph. Since there are $\overline{m}$ edges in $K[\overline{G}]$, there cannot be more than $2\overline{m}$ elements in $\mathscr{C}$. □

Now we are ready to explain the algorithm. The general algorithmic approach is similar to the one used in [1] but to make use of the structural insights that were discussed earlier we have to keep track of some additional data.

Linear Structure of Graphs and the Knotting Graph

- First of all we compute all components and all intervals of G and construct for each vertex x a list of the components in $G - N[x]$. Each vertex x is assigned an independence number $\alpha(x)$ that initially is set to zero and later contains the sum of the independence numbers of all those elements of $\mathscr{C}^x$ that have been processed already.
- We call components $C_1, C_2, \ldots, C_t$ *twins* if they contain the same set of vertices but occur at the deletion of the neighborhood of different vertices $x_1, x_2, \ldots, x_t$. We create a list of "interested pairs" of vertices, i.e., pairs of vertices x, y that have a common interest in the computation of the independence number of some component since this component is a twin for them. As we have not determined the twins yet, we simply select all pairs of non-adjacent vertices of G. Each of these pairs is assigned an independence number $\alpha(x, y)$, and again, initially it is set to zero. Later on it will contain the sum of the independence numbers of all those components that are twins for x and y and which have been processed already.
- Now we sort the components according to their size and within the set of components of the same size according to a lexicographic ordering. Using bucket sort, this can be done in order of the number of components of the whole graph (i.e., the number of components in $\mathscr{C}$) times the number of nodes per component. By Proposition 17 this is $O(\overline{m}n)$.
- Using this ordering we can identify twins. Let $C_1, C_2, \ldots, C_t$ be twins for a set of vertices $x_1, x_2, \ldots, x_t$. For those twins we leave only one copy C_1 in the list of all components of G and for C_1 we create a list of pointers to each of the vertices $x_1, \ldots, x_t$ that leave behind this component when removed together with the corresponding neighborhood. Whenever we have determined the independence number of C_1 we simply have to "inform" each of the vertices x_i and each of the interested pairs x_i, x_j, with $i \neq j, i, j \in \{1, \ldots, t\}$.
- If, during the computation, the value of the independence number of some component C is determined, this value is added both to the independence number of all vertices that have this component and to the independence number of all interested pairs, of this component. The updating of the independence numbers of the vertices takes $O(n^2)$ time since there are n vertices and each of the vertices has no more than n components. The updating of the independence numbers of the interested pairs takes $O(\overline{m}n)$ time since there are at most $\overline{m}$ interested pairs and each of those pairs has no more than n components in common.
- The computation of the independence number for a component is done in the same way as in the previous version of the algorithm, in $O(\overline{m})$.
- For computing the independence number of an interval $I(x, y)$, formula (5) is used. This is done as follows. For each interval $I(x, y)$ we have to consider for each $s \in I(x, y)$ the value of the independence number for the case that we select this vertex to be in the independent set. But instead of checking for each single component of $G - N[s]$ the corresponding independence number, we use the value $\alpha(s)$, which is the sum of the independence numbers of all components of $G - N[s]$ that have been computed already. Of course, no component of size equal or larger than the size of $I(x, y)$ can influence the independence number of $I(x, y)$ because it cannot be contained in $I(x, y)$. Only the independence number of components

that are contained in $I(x, y)$ have to be added to the value of the independence number. Hence, all values that are not yet computed are not of interest and the initial value zero is the right choice for this case. Of course, we might have added to $\alpha(s)$ also the independence number of components of $G - N[s]$ that have smaller size but are not contained in $I(x, y)$. For those components C we have two possible cases. Either C is one of $C^s(x)$ or $C^s(y)$. In this case the pointer $P(s, x)$ and $P(s, y)$ point to this component and we can easily determine the corresponding independence number and subtract this value from $\alpha(s)$. The second case seems to be more complicated. This is the case that we have added the independence number of some small component C of $G - N[s]$ to $\alpha(s)$ but this component is neither contained in $I(x, y)$ nor is it one of $C^s(x)$ and $C^s(y)$. But here our Theorem 15 helps. By this theorem, C has to be also a component either of $G - N[x]$ or of $G - N[y]$. The sum of the independence numbers of those components are stored in $\alpha(x, s)$ and $\alpha(s, y)$. Consequently, we can simply subtract this value from $\alpha(s)$, too. The only problem that still might occur is that there is a component C that is both a component for x and s and for y and s. The value of its independent number would, of course, be subtracted twice if we subtract both $\alpha(x, s)$ and $\alpha(s, y)$, although, it would be added only once by $\alpha(s)$. The solution to this problem is given by Lemma 16 since this lemma shows that the sum of the independence number of all those components C is stored in $\alpha(x, y)$.

Consequently, the value of the independence number of an interval $I(x, y)$ can be computed using a simple formula:

$$\alpha(I(x, y)) = 1 + \max_{s \in I(x,y)} (\alpha(I(x, s)) + \alpha(I(s, y)) \\ + \alpha(s) - P(s, x) - P(s, y) \\ - \alpha(x, s) - \alpha(s, y) + \alpha(x, y))$$

Altogether this leads to the following result.

Theorem 18 *There is an $O(n\overline{m})$ algorithm to compute the independence number of a given AT-free graph.*

Broersma et al. [1] showed that their algorithm can be extended to work for the weighted case of the problem as well, i.e., non-negative weights are assigned to the vertices and one searches for the maximum weight of an independent set. Using the same method, the above algorithm can also be modified to solve the weighted case of the problem within the same time bound.

5 Conclusions

As shown in this paper studying the linear structure of graphs is helpful for better understanding the properties of graph classes but also for efficiently recognizing them and for solving optimization problems. Motivated by the usefulness of the linear structure various other approaches for studying linearity have been suggested. Since cocomparability graphs have a simple characterization using linear orderings that are implied by the underlying partial order, people have also searched for linear orderings characterizing AT-free graphs. Using an altered version of the knotting graph such characterizations could be found for two subclasses of AT-free graphs [2]; recently also for AT-free graphs such a characterization has been proven [4]. Now the interesting question is whether this characterizing ordering can also be applied for solving optimization problems of this class of graphs. Good candidates for such optimization problems are the Hamiltonian path and cycle as well as the minimum coloring problem, which both can be solved in polynomial time on cocomparability graphs but are still open for the class of AT-free graphs.

References

1. Broersma, H., Kloks, T., Kratsch, D., Müller, H.: Independent sets in asteroidal triple-free graphs. SIAM J. Discret. Math. **12**(2), 276–287 (1999)
2. Corneil, D.G., Köhler, E., Olariu, S., Stewart, L.: Linear orderings of subfamilies of AT-free graphs. SIAM J. Discret. Math. **20**(1), 105–118 (2006)
3. Corneil, D.G., Olariu, S., Stewart, L.: Asteroidal triple-free graphs. SIAM J. Discret. Math. **10**(3), 399–430 (1997)
4. Corneil, D.G., Stacho, J.: Vertex ordering characterizations of graphs of bounded asteroidal number. J. Graph Theory **78**(1), 61–79 (2015)
5. Gallai, T.: Transitiv orientierbare graphen. Acta Math. Acad. Sci. Hung. **18**, 25–66 (1967)
6. Gilmore, P., Hoffman, A.: A characterization of comparability graphs and of interval graphs. Can. J. Math. **16**, 539–548 (1964)
7. Kelly, D.: Comparability graphs. In: Rival, I. (ed.) Graphs and Order, pp. 3–40. D. Reidel Publishing Company, Dordrecht (1985)
8. Köhler, E.: Recognizing graphs without asteroidal triples. J. Discret. Algorithms **2**(4), 439–452 (2004)
9. Köhler, E., Mouatadid, L.: A linear time algorithm to compute a maximum weighted independent set on cocomparability graphs. Submitted (2015)
10. McConnell, R.M., Spinrad, J.P.: Modular decomposition and transitive orientation. Discret. Math. **201**(1), 189–241 (1999)
11. Möhring, R.H.: Triangulating graphs without asteroidal triples. Discret. Appl. Math. **64**(3), 281–287 (1996)
12. Parra, A., Scheffler, P.: Characterizations and algorithmic applications of chordal graph embeddings. Discret. Appl. Math. **79**(1–3), 171–188 (1997)

Finding Longest Geometric Tours

Sándor P. Fekete

Abstract We discuss the problem of finding a longest tour for a set of points in a geometric space. In particular, we show that a longest tour for a set of n points in the plane can be computed in time $O(n)$ if distances are determined by the Manhattan metric, while the same problem is NP-hard for points on a sphere under Euclidean distances.

1 Introduction: Short and Long Roundtrips

The Traveling Salesman Problem (TSP) is one of the classic problems of combinatorial optimization. Given a complete graph $G = (V, E)$ with edge weights $c(e)$ for all edges $e \in E$, find a shortest roundtrip through all vertices, i.e., a cyclic permutation π from the symmetric group S_n of all n vertices $v_1, \ldots, v_n$, such that the total tour length $\sum_{i=1}^{n} c(\{v_i, v_{\pi(i)}\})$ is minimized.

The difficulties of finding a good roundtrip are well known. The classical Odyssey is illustrated in Fig. 1: according to legend, it took Ulysses many years to complete his voyage. One justification is the computational complexity of the TSP: it is one of the most famous NP-hard problems, so it does indeed take many years of CPU time to find provably optimal solutions for non-trivial instances.

However, there is an even more convincing justification for Ulysses' failure to be home in a more timely fashion: it was not him who chose his route. Instead, malevolent gods caused a deliberately long voyage—so the real objective was to *maximize* the traveled distance. This motivates the MaxTSP: Find a roundtrip that visits all vertices in a weighted graph, such that the total tour length is maximized.

In this chapter, we study longest tours in a geometric setting, in which the vertices are points in two- or three-dimensional space, and the edge weights are induced by the distance between them. As it turns out, the difficulty of the corresponding MaxTSP depends greatly on the involved geometry.

S.P. Fekete (✉)
Department Informatik, Technische Universität Braunschweig, Mühlenpfordtstraße 23, 38106 Braunschweig, Germany
e-mail: s.fekete@tu-bs.de

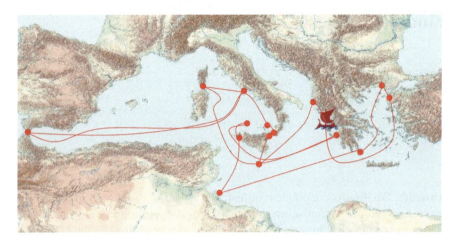

Fig. 1 The Odyssey: a tour through 16 locations in the Mediterranean

2 Traveling in Manhattan

The cost of traveling in a geometric space is measured by the geometric distance between points. A particularly simple way is to measure the axis-parallel distances separately, as one does when traveling along streets and avenues in Manhattan, giving rise to L_1 or *Manhattan distances*. A natural alternative is to use L_2 or *Euclidean distances*, which correspond to the length of straightline connections. When trying to find a shortest tour, this distinction does not make a difference in terms of the resulting problem complexity.

Theorem 1 *It is NP-complete to decide whether a set of n distinct points in the integer planar grid allows a roundtrip of length n.*

This amounts to deciding whether a *grid graph* has a Hamiltonian cycle; see Fig. 2. The corresponding distance is the same in Manhattan or Euclidean distances.

In the following we sketch why it is considerably easier to find a *longest* tour for a planar point set with Manhattan distances.

Theorem 2 *When distances are measured according to the Manhattan metric, finding a longest roundtrip for a set P of n points in the plane can be achieved in time $O(n)$.*

One key idea is to consider a *Manhattan median* $c = (x_c, y_c)$ for P, i.e., a point for which x_c is a median of all x-coordinates, and y_c is a median of all y-coordinates. Because c minimizes both the sum of all x- and y-distances to points in P, it induces a *Minimum Steiner Star*, as follows; we write $L_1(p, q)$ for the Manhattan distance between p and q.

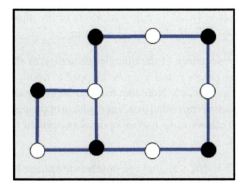

Fig. 2 A set of integer grid points induces a *grid graph*, in which two vertices are adjacent if and only if they have distance 1. Grid graphs are bipartite, as we can split the set of vertices into those with odd coordinate sum (*black*) and even coordinate sum (*white*). It is NP-complete to decide whether a given grid graph with n vertices has a tour of length n, i.e., a Hamiltonian cycle.

Lemma 3 *A Manhattan median c minimizes the total distance $\sum_{i=1}^{n} L_1(c, p_i)$ to all points p_i in P.*

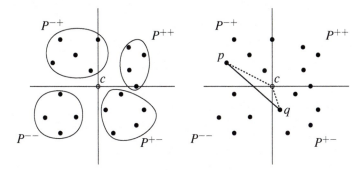

Fig. 3 (*Left*) A two-dimensional median c for a planar point set splits it into four quadrants, with equal numbers of points in opposite quadrants. (*Right*) In Manhattan distances, connecting two points in opposite quadrants incurs the same cost as connecting the median to both of them.

Finding c is possible in linear time. Because a median splits P into two equal subsets according to either x- or y-distances, it follows that it induces a split into four quadrants, such that there is an equal number of points in opposite quadrants; see Fig. 3(left). Furthermore, the weight of the corresponding Steiner Star, i.e., the sum of all distances from the median, induces an upper bound on the length of a longest tour.

Lemma 4 *For a set $P = \{p_1, \ldots, p_n\}$ of n points in the plane, we have the dual relationship* $\max_{\pi \in S_n^{cyclic}} \sum_{i=1}^n L_1(p_i, p_{\pi(i)}) \leq 2 \min_{c \in \mathbb{R}^2} \sum_{i=1}^n L_1(c, p_i)$.

This is a simple consequence of the triangle inequality, as shown in Fig. 3(right): each edge between two points p and q can be mapped to a path via a third point c, so $L_1(p, q) \leq L_1(p, c) + L_1(c, q)$. Note that for Manhattan distances, this inequality actually holds with *equality*, provided that p and q lie in opposite quadrants respective to c. This observation allows us to find an optimal permutation that consists of *two* cycles instead of one.

Lemma 5 *For a set $P = \{p_1, \ldots, p_n\}$ of n points in the plane, there is a permutation $\overline{\pi}$ consisting of two cycles for which* $\sum_{i=1}^n L_1(p_i, p_{\overline{\pi}(i)}) = 2 \min_{c \in \mathbb{R}^2} \sum_{i=1}^n L_1(c, p_i)$.

As shown in Fig. 4(left), such a solution consists of two subtours: cycles that go back and forth between opposite quadrants, which is possible because of the equal number of points in those quadrants.

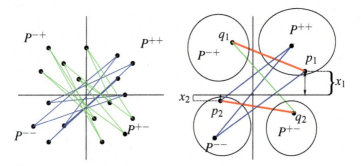

Fig. 4 (*Left*) Because opposite quadrants contain equal numbers of points, traveling back and forth between quadrants induces an optimal pair of subtours, shown in *blue* and *green*. (*Right*) In a tour, there must be connections between adjacent quadrants (shown in *red*), inducing an adjustment of the upper bound.

There is one final step left for obtaining an optimal tour. Observe that in order to form one connected cycle, any tour must contain a pair of edges (p_1, q_1) and (p_2, q_2) that connect *adjacent* quadrants, as shown in Fig. 4(right). This causes a gap in the triangle inequality, depending on the distance to a coordinate axis; in the example, this works out to $L_1(p_1, q_1) + 2|x_1| = L_1(p_1, c) + L_1(c, q_1)$ and $L_1(p_2, q_2) + 2|x_2| = L_1(p_2, c) + L_1(c, q_2)$. As a consequence, we must adjust the upper bound of $\sum_{i=1}^n L_1(c, p_i)$ by subtracting twice the smallest possible coordinate distances for one point from each of two opposite coordinate halfplanes, i.e., $2|x_1| + 2|x_2|$ in the example. Finding such a pair is easily possible in linear time. Now the corresponding pair of edges (p_1, q_1) and (p_2, q_2) can be used for merging both subtours into one tour that meets this upper bound, meaning that it is optimal.

Theorem 2 has a very powerful generalization for *polyhedral norms*, which are induced by using a symmetric convex polyhedron as the unit ball.

Theorem 6 *When distances are measured according to a polyhedral norm with a fixed number f of facets, finding a longest roundtrip for a set P of n points in a d-dimensional space for some fixed d can be achieved in time $O(n^{f-2} \log n)$.*

3 Traveling Around the Globe

In a graph setting, the TSP is NP-complete, and it is not difficult to see that this is also the case for the MaxTSP: If M is a sufficiently large number, replacing each edge weight $c(e)$ of a TSP instance by $M - c(e)$ yields a MaxTSP instance with the same optimal tours. But how can we use this simple idea of inverting short and long edges in a geometric setting? In fact, Theorem 2 indicates that this may not even be possible.

The key idea lies in switching from the plane to a sphere, and to consider Euclidean distances. Consider Fig. 5, which displays a sign posted at the airport of Auckland in New Zealand: It shows that furthest distances are to London and Frankfurt, which are both almost antipodal to the current location, i.e., close to the theoretical maximum distance of 20.000 km. Moreover, both cities are far from Auckland, but close to each other.

Utilizing this observation, we can conclude the following.

Theorem 7 *When distances are measured according to the Euclidean metric, finding a longest roundtrip for a set P of n points on a sphere is an NP-hard problem.*

We sketch a reduction from the Hamiltonicity of grid graphs, i.e., a consequence of Theorem 1. Let P be a set of planar distinct grid points, like the one shown in Fig. 2. We embed two "small" copies of P at antipodal locations of a sphere, as indicated in Fig. 6(left). This leaves corresponding pairs at maximum possible distance, as shown in Fig. 6(middle). Now observe that grid graphs are bipartite: each grid point has either even or odd sum of coordinates, and moving to an adjacent grid point changes parity. Omitting all *even* points of P from one of the two locations, but all *odd* points from the other (as depicted in Fig. 6(right)) leaves maximum possible distances between points that are almost antipodal (like Auckland and Frankfurt)—meaning that they are adjacent in the original grid graph. Therefore, there is a tour that uses only edges of this maximum possible length, if and only if the original grid graph has a Hamiltonian cycle. (The complete proof requires more involved trigonometry for a full argument, but that is a mere matter of math.)

Fig. 5 Large distances around the globe: a picture taken at Auckland airport, which is close to being 20.000 km away from, i.e., almost antipodal to, London and Frankfurt.

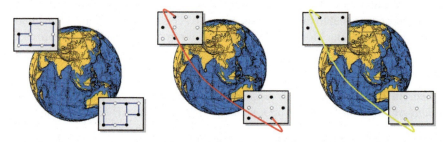

Fig. 6 Showing NP-hardness of the MaxTSP for points in 3D and Euclidean distances. (*Left*) Embedding two copies of a given grid graph G, such that corresponding vertices become antipodal points. (*Middle*) Longest distances within the resulting point set connect antipodal points. (*Right*) Exploiting bipartiteness of grid graphs for mapping edges in the original grid graph to longest edges in the remaining point set.

How hard is it to find a longest tour for Euclidean distances in the plane? This has been unresolved for more than a decade: it is Problem #49 of THE OPEN PROBLEMS PROJECT, and has been a challenge since 2003.

Problem 8 What is the complexity of finding a longest roundtrip for a set P of n points in the plane, when distances are measured according to the Euclidean metric?

4 Further Reading

There are different books describing various aspects of the TSP. A classic overview is provided by Lawler et al. [18]. A detailed exposition of the computational aspects involved in solving instances to optimality was presented by Applegate et al. [1], while Cook [7] gives an entertaining survey with various historical and anecdotal notes. The Odyssey instance was first presented by Grötschel and Padberg [14–16] and is contained in the benchmark library TSPLIB [21].

Papdimitriou [20] was the first to prove NP-hardness of the TSP for Euclidean distances in the plane; the NP-hardness result for Hamiltonicity in grid graphs is due to Itai et al. [17].

A couple of years before Arora [3] and Mitchell [19] independently showed that the geometric TSP can be approximated arbitrarily well (i.e., for any $\varepsilon > 0$, there is a polynomial-time algorithm that computes a tour within a factor of $(1 + \varepsilon)$ of the optimum), Barvinok [4] already did the same for the MaxTSP. Independently, Serdyuokov [22–24] gave several results for the MaxTSP. The bottleneck version of the MaxTSP (find a tour with a shortest edge that is as long as possible) was considered by Arkin et al. [2].

Barvinok et al. [5] were the first to provide a polynomial-time algorithm for the MaxTSP with polyhedral norms in fixed-dimensional space, as stated in Theorem 6. The strong geometric duality between Minimum Steiner Star and Maximum Matching was first observed by Tamir and Mitchell [25]. Fekete and Meijer [10, 11] proved a tight bound for the corresponding ratio between Minimum Steiner Star and Maximum Matching for Euclidean distances, and demonstrated in [12, 13] that this can be exploited for finding good maximum-weight matchings.

The results of this chapter (in particular, Theorems 2 and 7) were first presented in [9]. A more detailed journal version can be found in [6], which also contains full details of the conference paper [5].

See [8] for The Open Problems Project.

References

1. Applegate, D.L., Bixby, R.E., Chvátal, V., Cook, W.J.: The Traveling Salesman Problem: A Computational Study (Princeton Series in Applied Mathematics). Princeton University Press, Princeton (2007)
2. Arkin, E.M., Chiang, Y.J., Mitchell, J.S.B., Skiena, S., Yang, T.-C.: On the maximum scatter TSP. In: Proceedings of the 8th ACM-SIAM Symposium on Discrete Algorithms (SODA 97), pp. 211–220 (1997)
3. Arora, S.: Polynomial-time approximation schemes for Euclidean TSP and other geometric problems. J. ACM **45**(5), 753–782 (1998)
4. Barvinok, A.I.: Two algorithmic results for the Traveling Salesman Problem. Math. Oper. Res. **21**(1), 65–84 (1996)
5. Barvinok, A., Johnson, D.S., Woeginger, G.J., Woodroofe, R.: The maximum Traveling Salesman Problem under polyhedral norms. In: Proceedings of the 6th International Integer Pro-

gramming and Combinatorial Optimization Conference (IPCO VI). Lecture Notes in Computer Science, vol. 1412, pp. 195–201. Springer (1998)
6. Barvinok, A.I., Fekete, S.P., Johnson, D.S., Tamir, A., Woeginger, G.J., Wodroofe, R.: The geometric maximum Traveling Salesman Problem. J. ACM **50**, 641–664 (2003)
7. Cook, W.J.: In Pursuit of the Traveling Salesman: Mathematics at the Limits of Computation. Princeton University Press, Princeton (2011)
8. Demaine, E.D., Mitchell, J.S.B., O'Rourke, J.: The Open Problems Project (2001). http://cs.smith.edu/~orourke/TOPP/
9. Fekete, S.P.: Simplicity and hardness of the maximum Traveling Salesman Problem under geometric distances. In: Proceedings of the 10th ACM-SIAM Symposium Discrete Algorithms (SODA 99), pp. 337–345 (1999)
10. Fekete, S.P., Meijer, H.: On minimum stars, minimum Steiner stars, and maximum matchings. In: Proceedings of the 15th Annual ACM Symposium on Computational Geometry (SoCG 99), pp. 217–226. ACM (1999)
11. Fekete, S.P., Meijer, H.: On minimum stars and maximum matchings. Discrete Comput. Geom. **23**(3), 389–407 (2000)
12. Fekete, S.P., Meijer, H., Rohe, A., Tietze, W.: Solving a "hard" problem to approximate an "easy" one: Heuristics for maximum matchings and maximum Traveling Salesman Problems. In: Proceedings of the 3rd International Workshop on Algorithms Engineering and Experiments (ALENEX 2001). Lecture Notes in Computer Science, vol. 2153, pp. 1–16. Springer (2001)
13. Fekete, S.P., Meijer, H., Rohe, A., Tietze, W.: Solving a "hard" problem to approximate an "easy" one: Heuristics for maximum matchings and maximum Traveling Salesman Problems. J. Exp. Algorithms **7**, 21 (2002)
14. Grötschel, M., Padberg, M.W.: Ulysses 2000: In search of optimal solutions to hard combinatorial problems. SC 93-34, Zuse Institute Berlin, November 1993
15. Grötschel, M., Padberg, M.W.: Die optimierte Odyssee. Spektrum der Wiss. Dig. **2**, 32–41 (1999)
16. Grötschel, M., Padberg, M.W.: The optimized Odyssey. AIROnews **VI**(2), 1–7 (2001)
17. Itai, A., Papadimitriou, C.H., Swarcfiter, J.L.: Hamilton paths in grid graphs. SIAM J. Comput. **11**, 676–686 (1982)
18. Lawler, E.L., Lenstra, J.K., Rinnooy Kan, A.H.G., Shmoys, D.B. (eds.).: The Traveling Salesman Problem: A Guided Tour of Combinatorial Optimization. Wiley, Chichester (1985)
19. Mitchell, J.S.B.: Guillotine subdivisions approximate polygonal subdivisions: A simple polynomial-time approximation scheme for geometric TSP, k-MST, and related problems. SIAM J. Comput. **28**, 1298–1309 (1999)
20. Papadimitriou, C.H.: The Euclidean Traveling Salesman Problem is NP-complete. Theor. Comput. Sci. **4**, 237–244 (1977)
21. Reinelt, G.: TSPlib—A Traveling Salesman Problem library. ORSA J. Comput. **3**(4), 376–384 (1991)
22. Serdyukov, A.I.: An asymptotically exact algorithm for the Traveling Salesman Problem for a maximum in Euclidean space (Russian). Upravlyaemye Sistemy **27**, 79–87 (1987)
23. Serdyukov, A.I.: Asymptotic properties of optimal solutions of extremal permutation problems in finite-dimensional normed spaces (Russian). Metody Diskret Analiz **51**, 105–111 (1991)
24. Serdyukov, A.I.: The Traveling Salesman Problem for a maximum in finite-dimensional real spaces (Russian). Diskretnyi Analiz i Issledovanie Operatsii **2**(1), 50–56 (1995)
25. Tamir, A., Mitchell, J.S.B.: A maximum b-matching problem arising from median location models with applications to the roommates problem. Math. Program. **80**(2), 171–194 (1998)

Generalized Hanan Grids for Geometric Steiner Trees in Uniform Orientation Metrics

Matthias Müller-Hannemann

Abstract Given a finite set of points in some metric space, a fundamental task is to find a shortest network interconnecting all of them. The network may include additional points, so-called *Steiner points*, which can be inserted at arbitrary places in order to minimize the total length with respect to the given metric. This paper focuses on uniform orientation metrics where the edges of the network are restricted to lie within a given set of legal directions. We here review the crucial insight that many versions of geometric network design problems can be reduced to the Steiner tree problem in finite graphs, namely the *Hanan grid* or its extensions.

1 Introduction

The family of Steiner tree problems is among the most fundamental problems in combinatorial optimization. They have served as a testbed for many new algorithmic ideas—both in theory and in practice. Long ago, the interest in network design problems started within a geometric framework as a kind of facility location problem. Pierre de Fermat described in 1643 the most simple version as "given three points, a fourth is to be found, from which if three straight lines are drawn to the given points, the sum of the three lengths is minimum." What we nowadays call the Euclidean Steiner tree problem has first been posed by the French mathematician Joseph Diaz Gergonne in 1811: "A number of cities are located at known locations on a plane; the problem is to link them together by a system of canals whose total length is as small as possible." The optimal network can safely be assumed to be a tree since we can always delete the longest edge on each cycle without destroying connectivity.

Due to crucial applications in VLSI design, a strong interest arose in geometric versions where the orientation of edges in the network are restricted to vertical and horizontal directions (Manhattan style). Later this has been generalized to uniform

M. Müller-Hannemann (✉)
Institut für Informatik, Martin-Luther-Universität Halle-Wittenberg, Von-Seckendorfff-Platz 1, 06120 Halle (Saale), Germany
e-mail: muellerh@informatik.uni-halle.de

orientation metrics, where edge orientations are restricted to a small number of directions, uniformly separated by the same angle.

In order to shorten the length of the interconnecting network, it is allowed to introduce additional branching points, the so-called *Steiner points*. These Steiner points can—in principle—be placed everywhere. Hence, it is a priori unclear whether the location of Steiner points of minimum length networks can be restricted to a finite set. It is a beautiful result that many versions of geometric network design problems can be reduced in a simple way to combinatorial optimization problems on finite graphs, the Hanan grids or variations thereof. Given a connected graph $G = (V, E)$, a length function ℓ, and a set of terminals $K \subseteq V$, a *Steiner tree* is a connected, cycle-free subgraph of G containing all vertices of K. A Steiner tree T is a *Steiner minimum tree* of G if the length of T is minimum among all Steiner trees. The restriction to a finite solution space is a key achievement. It is a prerequisite for an algorithmic treatment of the problem (although all versions are NP-hard). One particular advantage is that it allows exact approaches like dynamic programming or branch-and-bound. Having transformed the geometric versions to pure graph problems, a rich arsenal of approaches for graphs becomes applicable. In two dimensions, the resulting graphs are planar. Hence, planarity can be exploited, which makes polynomial-time approximation schemes (PTAS) possible.

Overview. We start our discussion with rectilinear Steiner trees in two and higher dimensions, and sketch an application of the Hanan grid in the presence of obstacles (Sect. 2). Then we extend the basic ideas to more general metrics, the so-called uniform orientation metrics. We show that a more elaborate recursive construction of a generalized Hanan grid has the desired property to contain an optimal solution for the Steiner tree problem under these metrics (Sect. 3). Finally, we provide pointers to the literature (Sect. 4).

2 Rectilinear Steiner Trees

2.1 The Classical Case in Two Dimensions

The rectilinear Steiner tree problem is a key problem in VLSI layout. The basic version is defined as follows. We are given a finite set K of distinct points in the plane, the so-called *terminals*. The task is to construct a network, the so-called *rectilinear Steiner tree*, interconnecting all given terminals subject to the condition that all edges of the network have to be embedded only with horizontal and vertical segments. The length of a tree is the sum of the lengths of all its edges. A shortest rectilinear Steiner tree is called *rectilinear Steiner minimum tree*. See Fig. 1 for an example.

For rectilinear Steiner tree problems for point sets in the plane the most successful approaches are based on transformations to the related Steiner tree problem in graphs. Given a finite set K of points in the plane, the *Hanan grid* $G(K)$ is the graph

Fig. 1 Small examples of Steiner minimum trees in rectilinear and octilinear metric for the same set of given terminals in the plane

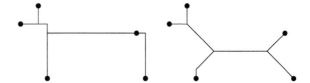

induced by constructing vertical and horizontal lines through each point in K. The intersections of these lines are called *grid points*, they form the vertex set of $G(K)$. Each grid point is connected by edges to all its direct neighbors on the given lines. The grid is named after Maurice Hanan who showed in a seminal paper in 1966 that the grid contains a rectilinear Steiner minimum tree.

Theorem 1 (Hanan 1966) *Given a finite set K of points in the plane, there exists a rectilinear Steiner minimum tree all of whose Steiner points are grid points of $G(K)$.*

We postpone the proof of this theorem to the next subsection, where we provide a simple and strikingly elegant proof for a slightly more general case which avoids lengthy case analyses.

2.2 Higher Dimensions

In dimension $d \geq 2$, edges of a rectilinear Steiner tree consist of contiguous line segments that run parallel to the d coordinate axes. Given a finite set K of points in $\mathbb{R}^d$, we extend our definition of the Hanan grid $G(K)$. For each point in K we take the d hyperplanes normal to the coordinate axes which run through it. The Hanan grid is then the grid induced by the union of all these hyperplanes. More precisely, its grid points are intersections of d mutually orthogonal hyperplanes from this set. Likewise, the edges are segments of straight lines, bounded by grid points, that are the intersection of $d - 1$ hyperplanes from this set.

The main result for rectilinear Steiner trees in d dimensions is that for any given finite point set $K \subset \mathbb{R}^d$, there exists a rectilinear Steiner minimum tree of K such that each of its Steiner points lies at an intersection point of the grid formed by hyperplanes normal to the coordinate axes.

Theorem 2 *Given a finite set K of points in $\mathbb{R}^d$, there exists a rectilinear Steiner minimum tree all of whose Steiner points are grid points of $G(K)$.*

Proof We call a coordinate *listed* if it appears as a coordinate of some point in K. So every grid point has only listed coordinates. Let T be a rectilinear Steiner minimum tree for K. Denote by U the set of unlisted coordinates occurring in Steiner points of T.

If U is empty, we are done. If U is non-empty, we show that we can always reduce the cardinality of U by one without increasing the length of the tree. Iterating

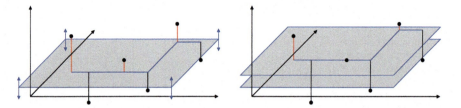

Fig. 2 Example of a three-dimensional Steiner tree with unlisted Steiner points on the highlighted hyperplane which can be slid up or down (*left*), and the modified tree of same length after shifting the hyperplane upwards (*right*)

this argument, we end up with a modified tree T' where all Steiner points are grid points of $G(K)$. So let $p = (p_1, p_2, \ldots, p_d)$ be a Steiner point in T with an unlisted coordinate p_i. Let H be the hyperplane passing through p that is normal to the x_i-axis. Since p_i is not listed, there is no terminal on this hyperplane. We can move this hyperplane with all edges on it along the x_i-axis, until it hits for the first time a listed or unlisted coordinate. See Fig. 2 for an example in three dimensions. Along with this move, all segments of T touching the hyperplane but oriented in direction parallel to the x_i-axis are reduced or enlarged, depending on the direction of the movement, so that the tree remains connected. Observe that the number of segments (parts of edges) of T touching the hyperplane on the two sides must be equal. (This also implies that the move is finite). If not, then we could move the hyperplane in direction of the side with the larger number of incident segments and thereby reduce the total length of the tree. This would be a contradiction to the assumption that T is a rectilinear Steiner minimum tree. Thus, moving the hyperplane does not change the length of T. Moreover, by such a move one unlisted coordinate vanishes, but no new unlisted coordinate has been created. □

2.3 Extensions

The idea of the Hanan grid construction can be generalized in several ways. We next elaborate on an application in VLSI design. To this end, let us come back to the Euclidean plane. An *obstacle* is a connected region in the plane bounded by one or more simple rectilinear polygons such that no two polygon edges have an inner point in common (i.e., an obstacle may contain holes). We require the obstacles to be disjoint, except for possibly a finite number of common points.

A classical task in VLSI design is to find the shortest rectilinear network interconnecting a set of points and avoiding the interior of all obstacles. This problem is called the *obstacle-avoiding Steiner minimum tree* problem.

The Hanan grid construction can be extended in the obvious way. In this context we consider the grid induced by a vertical and a horizontal line through each given point as before but augmented by a line through each edge used in the description

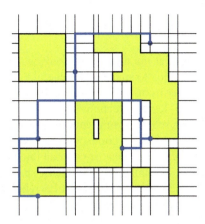

Fig. 3 The Hanan grid in the presence of rectilinear obstacles

of the obstacles. All grid points in the strict interior of obstacles and adjacent edges are removed; see Fig. 3 for an example. Given a finite point set K in the plane and a set of obstacles $\mathcal{O}$, we denote by $G(K, \mathcal{O})$ the corresponding grid.

Theorem 3 *Given an instance of the obstacle-avoiding Steiner minimum tree problem, specified by a point set K and a set of obstacles $\mathcal{O}$, there exists an obstacle-avoiding Steiner minimum tree all of whose Steiner points are grid points of $G(K, \mathcal{O})$.*

A formal proof of this theorem can be done along the lines of that of Theorem 2. The main idea is again to show that any optimal Steiner minimum tree can be transformed such that it lies completely on the Hanan grid. Figure 4 shows an example where the horizontal tree segment does not lie on the Hanan grid. If it is slid up or down until it hits the nearest vertical Hanan grid line, the tree length does not change but Steiner points with unlisted coordinates disappear. Further extensions of the classical Hanan grid will be mentioned in Sect. 4.

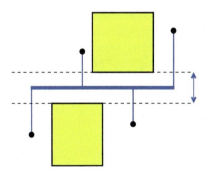

Fig. 4 A Steiner minimum tree not lying on the Hanan grid. Its bold horizontal segment, however, can be slid *up* or *down* until it hits one of the dashed Hanan grid lines.

3 Uniform Orientation Metrics

Rectilinear Steiner trees restrict the directions of edges to vertical and horizontal orientations. A strong motivation to study more general metrics than the traditional rectilinear routing also stems from VLSI design. For example, *octilinear routing* allows wiring in 45- and 135-degree directions in addition to vertical and horizontal wires. Clearly, such a routing scheme promises advantages in wire length reduction.

Therefore, a natural extension of the corresponding *Manhattan* or L_1-metric is to consider so-called *uniform orientation metrics*. For an integral parameter $\lambda \geq 2$, we consider λ legal orientations. The unit disk in this metric space, called the λ-*geometry* plane, is a centrally symmetric 2λ-gon with two vertices lying on the x-axis. See Fig. 5 for a few examples. Consecutive orientations are separated by a fixed angle of π/λ. A λ-geometry restricts every line segment to one of the given orientations $i\pi/\lambda$, for $0 \leq i < \lambda$, with respect to the positive x-axis.

A Steiner minimum tree in a λ-geometry is called λ-SMT. The *rectilinear* or *Manhattan* metric corresponds to the 2-geometry, the *hexagonal* metric to the 3-geometry, and the *octilinear* metric to the 4-geometry.

The orientation of angle $i\pi/\lambda$, for $0 \leq i < \lambda$, with the positive x-axis is referred to as the ith λ-*direction*. Two orientations separated by an angle of exactly π/λ are said to be *adjacent*. For any two points p and q in the plane, if the straight line segment connecting them is a legal direction, then the two points or the edge connecting them is said to be *straight*, otherwise they are *non-straight*. If p and q are non-straight, then an edge, representing the shortest way to connect them, consists of exactly two segments with orientations equal to two adjacent λ-directions. There are always two ways to lay out a non-straight edge between two non-straight points by two segments. These two possibilities form a parallelogram.

3.1 Basic Properties

Optimal Steiner trees in a λ-geometry have a number of important properties which we state without proof (we provide references in Sect. 4). The first of these properties assures that a Steiner tree does not require too many Steiner points. More precisely, an easy counting argument considering Steiner points of degree three or higher shows that $n - 2$ Steiner points always suffice. This bound is tight.

Fig. 5 The unit disks in λ-geometries for $\lambda = 2, 3,$ and 4, respectively

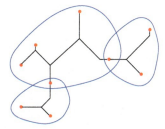

Fig. 6 An octilinear Steiner tree (the terminals in *red*) and its decomposition into full components

Lemma 4 *Given an instance with n terminals, at most $n - 2$ Steiner points appear in a Steiner minimum λ-tree.*

A straightforward case analysis also shows that Steiner points with degree higher than four must be locally suboptimal.

Lemma 5 *In a Steiner minimum λ-tree, the degree of any Steiner point is either 3 or 4. Moreover, all vertices adjacent to Steiner points of degree 4 are terminals.*

A Steiner minimum λ-tree is called *full* if all terminals are of degree one. Any Steiner λ-tree can be decomposed into its full components by splitting all its non-leaf terminals; see Fig. 6 for an example. A λ-tree is called *fulsome* if the number of its full components is maximized. Clearly, there is always a Steiner minimum λ-tree that is fulsome. Using a non-trivial shifting argument one can show the following crucial property.

Lemma 6 *There exists a fulsome Steiner minimum λ-tree such that at most one edge is non-straight in each of its full components.*

3.2 Multi-Level Hanan Grid

The obvious idea to generalize the Hanan grid to λ-geometries is to consider the grid induced by taking λ lines, one for each legal direction, through the given terminals. Unfortunately, this does not suffice. A small counterexample is shown in Fig. 7 for four terminals where the optimal octilinear Steiner tree requires a Steiner point which is not an intersection point of legal lines through the given terminals. We show next that a clever idea helps us to overcome these problems.

Du and Hwang generalized the Hanan grid construction to λ-geometries. They define *multi-level grids* $GG_k(K)$ recursively in the following way. For an instance with point set K, $GG_0(K) = K$. The grid $GG_1(K)$ is constructed by taking λ (infinite) lines with orientations $0, \pi/\lambda, 2\pi/\lambda, \ldots, (\lambda - 1)\pi/\lambda$ for each point of K. The kth grid $GG_k(K)$ for $k > 1$ is constructed from the $(k-1)$th grid by adding

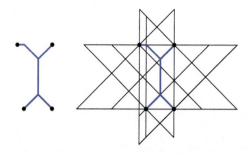

Fig. 7 A small example showing that an optimal Steiner tree in 4-geometry on four terminals may require a Steiner point (the *upper* one) that is not the intersection of legal directions through the terminals. Hence, GG_1 (shown on the *right*) does not suffice. However, since the lower Steiner point is a grid point of GG_1, the second level grid GG_2 includes the vertical segment of the Steiner tree and thereby also the upper Steiner point as the intersection of this vertical segment and the 45-degree line through the upper right terminal.

for each intersection point x of lines in $GG_{k-1}(K)$ additional lines through x with orientations $0, \pi/\lambda, 2\pi/\lambda, \ldots, (\lambda-1)\pi/\lambda$.

The main result about multi-level Hanan grids is a proof that GG_{n-2} contains all potentially necessary Steiner points for all $2 < \lambda < \infty$.

Theorem 7 *For all $2 < \lambda < \infty$ and for each set K of $n \geq 2$ points in the plane, there exists a Steiner minimum λ-tree T for K such that all Steiner points in T are grid points of GG_{n-2}.*

Proof Consider a full component T' of a fulsome Steiner minimum λ-tree for the given terminal set. Let K' be the terminals of T'. We have to show that all of its Steiner points are grid points of GG_{n-2}. By Lemma 6, this component has at most one non-straight edge. If T' has a Steiner point, then with the help of Lemma 5 we observe that there must be at least two terminals in T' which are adjacent to the same Steiner point, say p, and connected to it by straight edges. Hence, p lies on $GG_1(K')$. Each Steiner point adjacent to two vertices of $GG_{i-1}(K')$ with straight edges lies on $GG_i(K')$. Since we have at most $n-2$ Steiner points (Lemma 4), we see by induction that all of them lie on $GG_{n-2}(K')$. □

For some small values of λ the result can be significantly improved. When $\lambda = 3$ or $\lambda = 4$, one can show that for any point set K of n terminals there is a Steiner minimum λ-tree such that all Steiner points are represented in $GG_{\lceil (n-2)/2 \rceil}(K)$ or $GG_{\lceil 2n/3 \rceil - 1}(K)$, respectively. We omit the technically involved proofs of these results.

3.3 Approximations

We cannot neglect one striking drawback of multi-level grids: The graph $GG_k(K)$ has $\Theta(n^{2^k})$ vertices and edges. Hence, in general, if we want to guarantee an exact solution, we require an exponentially large graph.

In practice, we may therefore prefer approximate solutions. Since $GG_1(K)$ contains a shortest path between any pair of terminals, it also contains the solution obtained from the minimum spanning tree heuristic to approximate the minimum Steiner tree. Therefore, its performance guarantee cannot be worse than the Steiner ratio. The *Steiner ratio* is the smallest upper bound on the ratio between the length of a minimum spanning tree and the length of a Steiner minimum tree. To give a concrete example, the Steiner ratio in the octilinear case is $\frac{4}{2+\sqrt{2}}$. This implies that $GG_1(K)$ contains a solution which is at most about 17.15 % above the minimum for $\lambda = 4$.

There is a trade-off between the size of the constructed graph and the approximation guarantee. By appropriately refining the grid $GG_1(K)$ with additional lines parallel to the existing ones, one can achieve a $(1 + \varepsilon)$–approximation guarantee with a grid of size $O(n^2/\varepsilon^2)$ for every $\varepsilon > 0$.

4 Notes and Further Reading

The literature on Steiner tree problems is very comprehensive. For an introduction see, for example, the monographs by Hwang, Richards, and Winter [11] or Prömel and Steger [18]. The history of the Steiner tree problem, first considered in the Euclidean version, goes back to the French mathematician and logician Joseph Diaz Gergonne in the early nineteenth century, long before Jacob Steiner has worked on it [3]. The quotes of de Fermat and Gergonne, given in the introduction and translated from the original Latin and French, can be found in [7, p. 153] and [9], respectively.

The computational complexity of the Steiner tree problem has first been settled for the graph version. Karp showed that the Steiner tree problem in graphs is NP-hard, even for unit weights [12]. Bern and Plassmann showed that it is even MAXSNP-hard [1]. Restricted to planar graphs, the Steiner tree problem remains NP-hard [8], but has a polynomial-time approximation scheme, found by Borradaile, Klein, and Mathieu [2]. The NP-hardness of the rectilinear Steiner tree problem has been shown by Garey and Johnson [8]. Inspired by their proof, Müller-Hannemann and Schulze showed the NP-hardness of the octilinear Steiner tree problem [17]. Only very recently, Brazil and Zachariasen extended these NP-hardness results to all uniform orientation metrics [5].

The Hanan grid construction first appeared in a seminal paper by Maurice Hanan in 1966 [10]. The generalization of Hanan's original grid to d dimensions and the corresponding Theorem 2 has first been proved by Snyder [19]. The proof given here, however, follows closely the much simpler proof by Du and Hwang [6].

Zachariasen presented a catalogue of variants where other generalized versions of the Hanan grid can be applied [21]. These extensions include weighted regions, group Steiner trees, prize-collecting Steiner tree problems, minimum Manhattan networks, and applications in facility location. The extendability has, of course, limits. For example, if one considers soft obstacles, the Hanan property is destroyed [15, 16]. In a soft obstacle scenario the Steiner tree is allowed to run over it, but the length of the intersection of the tree with the obstacles is upper bounded by some maximal length L.

The multi-level Hanan grid theorem for λ-geometries has been presented by Lee and Shen [13] (where the proof has only been sketched for $\lambda = 4$). A rigorous proof can be found in a paper by Brazil et al. [4]. Yan et al. [20] carefully analysed the hexagonal metric ($\lambda = 3$) and improved the bound on the required level for the multi-level Hanan grid. They proved that all Steiner points are in $GG_{\lceil 2n/3 \rceil}$ for n given terminals. In the same spirit, Lin and Xue considered the octilinear case in more detail. They showed that all Steiner points are in $GG_{\lceil 2n/3 \rceil - 1}$ for n given terminals [14]. Bounds on the Steiner ratio for λ-geometries appear in [13]. The $(1+\varepsilon)$-approximation guarantee for a Steiner minimum λ-tree with a grid of size $O(n^2/\varepsilon^2)$ for every $\varepsilon > 0$ is from [16].

References

1. Bern, M.W., Plassmann, P.E.: The Steiner problem with edge lengths 1 and 2. Inf. Process. Lett. **32**(4), 171–176 (1989)
2. Borradaile, G., Klein, P.N., Mathieu, C.: An $O(n \log n)$ approximation scheme for Steiner tree in planar graphs. ACM Trans. Algorithm. **5**(3) (2009)
3. Brazil, M., Graham, R.L., Thomas, D.A., Zachariasen, M.: On the history of the Euclidean Steiner tree problem. Arch. Hist. Exact Sci. **68**, 327–354 (2014)
4. Brazil, M., Thomas, D., Winter, P.: Minimum networks in uniform orientation metrics. SIAM J. Comput. **30**(5), 1579–1593 (2000)
5. Brazil, M., Zachariasen, M.: The uniform orientation Steiner tree problem is NP-hard. Int. J. Comput. Geom. Appl. **24**(2), 87–106 (2014)
6. Du, D.Z., Hwang, F.: Reducing the Steiner problem in a normed space. SIAM J. Comput. **21**(6), 1001–1007 (1992)
7. de Fermat, P.: Oeuvres, vol. 1 (1891)
8. Garey, M., Johnson, D.: The rectilinear Steiner tree problem is NP-complete. SIAM J. Appl. Math. **32**, 826–834 (1977)
9. Gergonne, J.D.: Questions proposées. Problèmes de géométrie. Annales de Mathématiques pures et appliquées **1**, 292 (1810–1811)
10. Hanan, M.: On Steiner's problem with rectilinear distance. SIAM J. Appl. Math. **14**, 255–265 (1966)
11. Hwang, F.K., Richards, D.S., Winter, P.: The Steiner tree problem. Ann. Discrete Math. **53** (1992)
12. Karp, R.M.: Reducibility among combinatorial problems. In: Miller, R.E., Thatcher, J.W. (eds.) Complexity of Computer Computations, pp. 85–103. Plenum Press, New York (1972)
13. Lee, D.T., Shen, C.F.: The Steiner minimal tree problem in the λ-geometry plane. In: Proceedings 7th International Symposium on Algorithms and Computations (ISAAC 1996). Lecture Notes in Computer Science, vol. 1178, pp. 247–255. Springer (1996)

14. Lin, G.H., Xue, G.: Reducing the Steiner problem in four uniform orientations. Networks **35**(4), 287–301 (2000)
15. Müller-Hannemann, M., Peyer, S.: Approximation of rectilinear Steiner trees with length restrictions on obstacles. In: Dehne, F.K.H.A., Sack, J., Smid, M.H.M. (eds.) Algorithms and Data Structures, WADS 2003. Lecture Notes in Computer Science, vol. 2748, pp. 207–218. Springer (2003)
16. Müller-Hannemann, M., Schulze, A.: Approximation of octilinear Steiner trees constrained by hard and soft obstacles. In: Arge, L., Freivalds, R. (eds.) Algorithm Theory - SWAT 2006. Lecture Notes in Computer Science, vol. 4059, pp. 242–254. Springer (2006)
17. Müller-Hannemann, M., Schulze, A.: Hardness and approximation of octilinear Steiner trees. Int. J. Comput. Geom. Appl. **17**(3), 231–260 (2007)
18. Prömel, H., Steger, A.: The Steiner Tree Problem: A Tour through Graphs, Algorithms, and Complexity. Advanced Lectures in Mathematics, Vieweg (2002)
19. Snyder, T.L.: On the exact location of Steiner points in general dimension. SIAM J. Comput. **21**(1), 163–180 (1992)
20. Yan, G.Y., Albrecht, A.A., Young, G.H.F., Wong, C.K.: The Steiner tree problem in orientation metrics. J. Comput. Syst. Sci. **55**, 529–546 (1997)
21. Zachariasen, M.: A catalog of Hanan grid problems. Networks **38**, 76–83 (2001)

Budgeted Matching via the Gasoline Puzzle

Guido Schäfer

Abstract We consider a natural generalization of the classical matching problem: In the *budgeted matching problem* we are given an undirected graph with edge weights, non-negative edge costs and a budget. The goal is to compute a matching of maximum weight such that its cost does not exceed the budget. This problem is weakly NP-hard. We present the first polynomial-time approximation scheme for this problem. Our scheme computes two solutions to the Lagrangian relaxation of the problem and patches them together to obtain a near-optimal solution. In our patching procedure we crucially exploit the adjacency relations of vertices of the matching polytope and the solution to an old combinatorial puzzle.

1 Problem Definition

The *budgeted matching problem* is a natural generalization of the classical matching problem: We are given an undirected graph $G = (V, E)$ with edge weights $w : E \to \mathbb{Q}$, non-negative edge costs $c : E \to \mathbb{Q}^+$ and a budget $B \in \mathbb{Q}^+$. Recall that a matching of G is a subset $M \subseteq E$ of the edges such that no two edges of M share a common node. Let $\mathcal{F}$ be the set of all matchings of G. Define the weight of a matching M as the total weight of all edges in M, i.e., $w(M) := \sum_{e \in M} w(e)$. Similarly, the cost of M is defined as $c(M) := \sum_{e \in M} c(e)$. The goal is to compute a matching of maximum weight whose cost is at most B, i.e.,

$$\text{maximize } w(M) \text{ subject to } M \in \mathcal{F}, \; c(M) \leq B. \qquad (\bar{\Pi})$$

The exposition of the results given here is based on the article [1].

G. Schäfer (✉)
Centrum Wiskunde & Informatica, Science Park 123, 1098 XG Amsterdam, The Netherlands
e-mail: g.schaefer@cwi.nl

G. Schäfer
VU University Amsterdam, De Boelelaan 1105, 1081 HV Amsterdam, The Netherlands

© Springer International Publishing Switzerland 2015
A.S. Schulz et al. (eds.), *Gems of Combinatorial Optimization
and Graph Algorithms*, DOI 10.1007/978-3-319-24971-1_5

The budgeted matching problem is weakly NP-hard even for bipartite graphs. This follows by a simple reduction from the knapsack problem. Here we present the first polynomial-time approximation scheme (PTAS) for the budgeted matching problem. For a given $\varepsilon > 0$, our algorithm computes a $(1-\varepsilon)$-approximate solution to the problem in time $O(m^{O(1/\varepsilon)})$, where m is the number of edges in the graph.

Subsequently, we use OPT to refer to the weight of an optimal solution M^* to $(\bar{\Pi})$. Also, we use (Π) to refer to the respective unbudgeted matching problem (where the budget constraint "$c(M) \leq B$" is dropped).

2 A PTAS for Budgeted Matching

Consider the Lagrangian relaxation LR(λ) of the budgeted matching problem $(\bar{\Pi})$:

$$z(\lambda) := \text{maximize } \bigl(w(M) + \lambda(B - c(M))\bigr) \text{ subject to } M \in \mathcal{F}. \quad (\text{LR}(\lambda))$$

Note that every feasible solution to the budgeted matching problem $(\bar{\Pi})$ satisfies $c(M) \leq B$. Thus, for every $\lambda \geq 0$ the optimal solution to LR(λ) gives an upper bound on OPT, i.e., $z(\lambda) \geq \text{OPT}$. The Lagrangian dual problem is to find the best such upper bound, i.e., to determine λ^* such that $z(\lambda^*) = \min_{\lambda \geq 0} z(\lambda)$ (see also Fig. 1).

Note that for a fixed value of λ the Lagrangian relaxation LR(λ) is equivalent to solving a maximum weight matching problem with respect to the *Lagrangian weights*

$$w_\lambda(e) := w(e) - \lambda c(e) \quad \forall e \in E.$$

Given that there are combinatorial algorithms to solve the maximum weight matching problem, we can use standard parametric search techniques (see Sect. 4 for references) to determine an optimal Lagrangian multiplier λ^* in strongly polynomial time. In addition, we can compute within the same time bound two optimal matchings M_1 and M_2 to LR(λ^*) such that $c(M_1) \leq B \leq c(M_2)$.

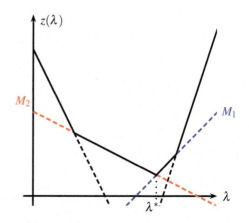

Fig. 1 The Lagrangian value $z(\lambda)$ as a function of λ (*solid line*). Each *dashed line* represents the Lagrangian value of a specific solution.

The idea now is to *patch* M_1 and M_2 together to obtain a feasible solution M to $(\bar{\Pi})$ whose weight $w(M)$ is not too far from the optimal one. More precisely, the following lemma will be crucial to derive our polynomial-time approximation scheme:

Lemma 1 (Patching Lemma) *There is a polynomial-time algorithm to compute a solution M to the budgeted matching problem of weight $w(M) \geq \text{OPT} - 2w_{\max}$, where $w_{\max}$ is the largest weight of an edge.*

A formal proof of this lemma is given in Sect. 3. Intuitively, our patching procedure consists of two phases: an *exchange phase* and an *augmentation phase*.

Exchange Phase: Consider the polytope induced by the set of feasible matchings $\mathcal{F}$ and let F be the face given by the solutions of maximum Lagrangian weight w_{λ^*}. This face contains both M_1 and M_2. We now iteratively replace either M_1 or M_2 with another vertex on F, preserving the invariant $c(M_1) \leq B \leq c(M_2)$, until M_1 and M_2 correspond to adjacent vertices of the matching polytope. Note that the Lagrangian weight of M_i, $i \in \{1, 2\}$, is $w_{\lambda^*}(M_i) = z(\lambda^*) \geq \text{OPT}$. However, with respect to the original weight, we can only infer that $w(M_i) = z(\lambda^*) - \lambda^*(B - c(M_i))$. That is, we cannot hope to use these matchings directly: M_1 is a feasible solution to $(\bar{\Pi})$ but its weight $w(M_1)$ might be arbitrarily far from OPT. In contrast, M_2 has weight $w(M_2) \geq \text{OPT}$, but is infeasible.

Augmentation Phase: In order to overcome the above problem, we exploit the adjacency relation between M_1 and M_2. It is known that two matchings M_1 and M_2 are adjacent in the matching polytope if and only if their symmetric difference $X = M_1 \oplus M_2$ is an alternating cycle or a path. The idea now is to patch M_1 according to a properly chosen subpath X' of X. We ensure that the subpath X' is chosen such that the Lagrangian weight of M_1 does not decrease too much, while at the same time the gap between the budget B and the cost of M_1 (and hence also the gap between $w(M_1)$ and $z(\lambda^*)$) is reduced. This way we obtain a feasible solution M whose weight differs from OPT by at most $2w_{\max}$.

Surprisingly, our proof that such a patching subpath X' always exists is based on the solution of an old combinatorial puzzle, also known as the *Gasoline Puzzle*:

> Along a speed track there are some gas-stations. The total amount of gasoline available in them is equal to what our car (which has a very large tank) needs for going around the track. Prove that there is a gas-station such that if we start there with an empty tank, we shall be able to go around the track without running out of gasoline.

With the help of our Patching Lemma we derive a polynomial-time approximation scheme by "guessing" the $\Theta(1/\varepsilon)$ largest weight edges in an optimum solution.

Theorem 1 *There is a deterministic algorithm that for every $\varepsilon > 0$ computes a solution to the budgeted matching problem of weight at least $(1 - \varepsilon)\text{OPT}$ in time $O(m^{2/\varepsilon + O(1)})$, where m is the number of edges in the graph.*

Proof Let $\varepsilon \in (0, 1)$ be a given constant. Assume that the optimum matching M^* contains at least $p := \lceil 2/\varepsilon \rceil$ edges. (Otherwise the problem can be solved optimally by complete enumeration.)

Consider the following algorithm: First, we guess the p largest weight edges M_H^* of M^*. We then remove from the graph G the edges in M_H^*, all edges incident to M_H^*, and all edges of weight larger than the smallest weight in M_H^*. We also decrease the budget by $c(M_H^*)$. Let I' be the resulting budgeted matching instance. Note that the maximum weight of an edge in I' is

$$w'_{\max} \le \tfrac{1}{p} w(M_H^*) \le \tfrac{1}{2} \varepsilon w(M_H^*).$$

Moreover, $M_L^* := M^* \setminus M_H^*$ is an optimum solution to I'. We then compute a matching M' for I' using the Patching Lemma and output the feasible solution $M := M_H^* \cup M'$.

For a given choice of M_H^* the running time of the algorithm is dominated by the time to compute the two solutions M_1 and M_2. This can be accomplished in $O(m^{O(1)})$ time using Megiddo's parametric search technique. Hence the overall running time of the algorithm is $O(m^{p+O(1)})$, where the m^p factor is due to the guessing of M_H^*.

By our Patching Lemma, $w(M') \ge w(M_L^*) - 2w'_{\max}$. It follows that

$$w(M) = w(M_H^*) + w(M') \ge w(M_H^*) + w(M_L^*) - 2w'_{\max}$$
$$\ge w(M^*) - \varepsilon w(M_H^*) \ge (1 - \varepsilon) w(M^*).$$

$\square$

3 Proof of the Patching Lemma

Let λ^* be the optimal Lagrangian multiplier and let M_1 and M_2 be two matchings of maximum Lagrangian weight $w_{\lambda^*}(M_1) = w_{\lambda^*}(M_2)$ such that $c(M_1) \le B \le c(M_2)$. Recall that M^* refers to an optimal solution to $(\bar{\Pi})$.

Observe that for $i \in \{1, 2\}$ we have that

$$w_{\lambda^*}(M_i) + \lambda^* B \ge w_{\lambda^*}(M^*) + \lambda^* B \ge w_{\lambda^*}(M^*) + \lambda^* c(M^*) = \text{OPT}. \quad (1)$$

Also note that by the optimality of M_1 and M_2, $w_{\lambda^*}(e) \ge 0$ for all $e \in M_1 \cup M_2$.

We next show how to extract from $M_1 \cup M_2$ a matching M with the desired properties in polynomial time. As outlined above, our patching procedure proceeds in two phases:

Exchange phase: Consider the symmetric difference $M' = M_1 \oplus M_2$. Recall that $M' \subseteq M_1 \cup M_2$ consists of a disjoint union of paths $\mathcal{P}$ and cycles $\mathcal{C}$. We apply the following procedure until eventually $|\mathcal{P} \cup \mathcal{C}| \le 1$: Take some $X \in \mathcal{P} \cup \mathcal{C}$ and let $A := M_1 \oplus X$. If $c(A) \le B$ replace M_1 by A. Otherwise replace M_2 by A. Note that this way we maintain the invariant $c(M_1) \le B \le c(M_2)$.

Note that in each step the number of connected components in $M_1 \oplus M_2$ decreases; hence this procedure terminates after at most $O(n)$ steps. Moreover, by the optimality of M_1 and M_2 the Lagrangian weight of the two matchings does not change during the process, i.e., the two matchings remain optimal. To see this note that if there is some $X \in \mathcal{P} \cup \mathcal{C}$ such that $w_{\lambda^*}(M_1 \oplus X) < w_{\lambda^*}(M_1)$ then there must exist some $X' \in \mathcal{P} \cup \mathcal{C}$ such that $w_{\lambda^*}(M_1 \oplus X') > w_{\lambda^*}(M_1)$, which is a contradiction to the optimality of M_1. It follows that $w_{\lambda^*}(A) = w_{\lambda^*}(M_1) = w_{\lambda^*}(M_2)$.

Note that at the end of this phase we have for every $i \in \{1, 2\}$

$$w(M_i) = w_{\lambda^*}(M_i) + \lambda^* c(M_i) = w_{\lambda^*}(M_i) + \lambda^* B - \lambda^*(B - c(M_i))$$
$$\geq \text{OPT} - \lambda^*(B - c(M_i)), \qquad (2)$$

where the inequality follows from (1).

In particular, if $c(M_i) = B$ for some $i \in \{1, 2\}$, we are done: M_i is a feasible solution to the budgeted matching problem and $w(M_i) \geq \text{OPT}$. Otherwise, we continue with the augmentation phase.

Augmentation Phase: The symmetric difference $M_1 \oplus M_2$ now consists of a unique path or cycle

$$X = (x_0, x_1, \ldots, x_{k-1}) \subseteq E$$

such that

$$c(M_1 \oplus X) = c(M_2) > B > c(M_1).$$

Observe that from (2) it follows that M_1 is a feasible solution whose original weight is close to optimal if its cost is sufficiently close to the budget B. The basic idea is to exchange edges along a subpath X' of X in order to obtain a feasible solution whose cost is close to the budget but still has large Lagrangian weight.

To this aim, we exploit the Gasoline Lemma. A formal statement is given below. We leave the proof of this lemma to the reader.

Lemma 2 (Gasoline Lemma) *Let $a_0, \ldots, a_{k-1}$ be a sequence of k real numbers such that $\sum_{j=0}^{k-1} a_j = 0$. There exists an index $i \in \{0, \ldots, k-1\}$ such that for every $h \in \{0, \ldots, k-1\}$,*

$$\sum_{j=i}^{i+h} a_{j \pmod{k}} \geq 0.$$

Consider the sequence

$$a_0 = \delta(x_0) w_{\lambda^*}(x_0), \quad a_1 = \delta(x_1) w_{\lambda^*}(x_1), \quad \ldots, \quad a_{k-1} = \delta(x_{k-1}) w_{\lambda^*}(x_{k-1}),$$

where $\delta(x_i) = 1$ if $x_i \in M_2$ and $\delta(x_i) = -1$ otherwise. (Note that, if X is a path, x_0 and x_{k-1} might both belong to either M_1 or M_2.) This sequence has total value

$\sum_{j=0}^{k-1} a_j = 0$ because of the optimality of M_1 and M_2. By the Gasoline Lemma there exists an edge x_i, $i \in \{0, 1, \ldots, k-1\}$, of X such that for any cyclic subsequence

$$X' = (x_i, x_{(i+1) \bmod k}, \ldots, x_{(i+h) \bmod k}),$$

where $h \in \{0, \ldots, k-1\}$, we have that

$$0 \le \sum_{j=i}^{i+h} a_{j \bmod k} = \sum_{e \in X' \cap M_2} w_{\lambda^*}(e) - \sum_{e \in X' \cap M_1} w_{\lambda^*}(e). \tag{3}$$

Let X' be the longest such subsequence satisfying $c(M_1 \oplus X') \le B$ (see Fig. 2 for examples). Note that X' consists of either one or two alternating paths. (The latter case only occurs if X is a path whose first and last edge belong to X'.) Let $e_1 = x_i$. Without loss of generality, we can assume $e_1 \in M_2$. (X' might start with one or two edges of M_1 with Lagrangian weight zero, in which case the next edge in M_2 is a feasible starting point of X' as well.)

Observe that $M_1 \oplus X'$ is not a matching unless X is a path and e_1 its first edge. However, $M := (M_1 \oplus X') \setminus \{e_1\}$ is always a matching. Moreover,

$$c(M) = c(M_1 \oplus X') - c(e_1) \le c(M_1 \oplus X') \le B.$$

That is, M is a feasible solution to the budgeted matching problem.

It remains to lower bound the weight of M. We have

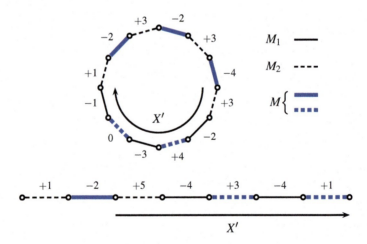

Fig. 2 Examples illustrating the construction used in the proof of the Patching Lemma. Each edge x_i is labeled with the value a_i.

$$w(M_1 \oplus X') = w_{\lambda^*}(M_1 \oplus X') + \lambda^* c(M_1 \oplus X')$$
$$= w_{\lambda^*}(M_1 \oplus X') + \lambda^* B - \lambda^*(B - c(M_1 \oplus X'))$$
$$\geq w_{\lambda^*}(M_1) + \lambda^* B - \lambda^*(B - c(M_1 \oplus X'))$$
$$\geq \text{OPT} - \lambda^*(B - c(M_1 \oplus X')), \quad (4)$$

where the first inequality follows from (3) and the second inequality follows from (1).

Let $e_2 = x_{(i+h+1) \pmod{k}}$. The maximality of X' implies that $c(e_2) > B - c(M_1 \oplus X') \geq 0$. Moreover, by the optimality of M_1 and M_2, the Lagrangian weight of any edge $e \in M_1 \cup M_2$ is non-negative, and thus $0 \leq w_{\lambda^*}(e_2) = w(e_2) - \lambda^* c(e_2)$. Altogether

$$\lambda^*(B - c(M_1 \oplus X')) \leq \lambda^* c(e_2) \leq w(e_2)$$

and hence by (4)

$$w(M_1 \oplus X') \geq \text{OPT} - w(e_2).$$

We conclude that

$$w(M) = w(M_1 \oplus X') - w(e_1) \geq \text{OPT} - w(e_2) - w(e_1) \geq \text{OPT} - 2w_{\max},$$

which proves the Patching Lemma. □

4 Extension and Notes on the Literature

The presented polynomial-time approximation scheme also extends to the *budgeted matroid intersection problem*. Here, we are given two matroids $\mathcal{M}_1 = (E, \mathcal{F}_1)$ and $\mathcal{M}_2 = (E, \mathcal{F}_2)$ on a common ground set of elements E. (We assume that these matroids are given implicitly by an *independence oracle*.) Moreover, we are given element weights $w : E \to \mathbb{Q}$, element costs $c : E \to \mathbb{Q}^+$ and a budget $B \in \mathbb{Q}^+$. The set of all feasible solutions $\mathcal{F} := \mathcal{F}_1 \cap \mathcal{F}_2$ is defined by the intersection of $\mathcal{M}_1$ and $\mathcal{M}_2$. The weight of an independent set $X \in \mathcal{F}$ is defined as $w(X) := \sum_{e \in X} w(e)$ and the cost of X is $c(X) := \sum_{e \in X} c(e)$. The goal is to compute a common independent set $X^* \in \mathcal{F}$ of maximum weight $w(X^*)$ among all feasible solutions $X \in \mathcal{F}$ satisfying $c(X) \leq B$.

Problems that can be formulated as the intersection of two matroids are, for example, matchings in bipartite graphs, arborescences in directed graphs, spanning forests in undirected graphs, etc. Although technically more involved, the ideas underlying our polynomial-time approximation scheme for the budgeted matching problem extend to this problem. More details can be found in [1].

Our algorithm needs to compute an optimal Lagrangian multiplier λ^* together with two respective optimal solutions. This can be done in polynomial time by standard techniques whenever the unbudgeted problem (Π) can be solved in polynomial time [9]. It can even be done in strongly polynomial time by using Megiddo's parametric search technique [4]. This technique can be used because *combinatorial* algorithms (only using comparisons and additions of weights) exist for (Π) (see, e.g., [10]). A similar idea was used by Goemans and Ravi [8] to derive a strongly polynomial-time approximation scheme for the constrained minimum spanning tree problem.

The description of the Gasoline Puzzle is taken from the book "Combinatorial Problems and Exercises" by Lovász [3, Problem 3.21].

Naor et al. [6] proposed a fully polynomial-time approximation scheme (FPTAS) for a rather general class of problems, which contains the budgeted matching problem considered here as a special case. However, personal communication revealed that, unfortunately, the stated result [6, Theorem 2.2] is incorrect.

An interesting open problem is whether there is a fully polynomial-time approximation scheme for the budgeted matching problem. We conjecture that budgeted matching is not strongly NP-hard. However, finding an FPTAS for this problem might be a very difficult task because of its relation to the *exact perfect matching problem*: In this problem, we are given an undirected graph $G = (V, E)$ with edge weights $w : E \to \mathbb{Q}$ and a parameter $W \in \mathbb{Q}$. The goal is to find a perfect matching of weight exactly W (if it exists).

This problem was first posed in 1982 by Papadimitriou and Yannakakis [7]. The problem admits a polynomial-time Monte Carlo algorithm [2, 5] if the edge weights are polynomially bounded. It is thus very unlikely that the exact perfect matching problem with polynomial weights is NP-hard because this would imply that RP = NP. However, the problem of finding a deterministic algorithm to solve the exact perfect matching problem has remained open so far. For polynomial weights and costs the budgeted matching problem is equivalent to the exact perfect matching problem; see [1] for more details. As a consequence, a (deterministic) FPTAS for the budgeted matching problem would resolve a long-standing open problem.

References

1. Berger, A., Bonifaci, V., Grandoni, F., Schäfer, G.: Budgeted matching and budgeted matroid intersection via the gasoline puzzle. Math. Program. **128**(1–2), 355–372 (2011)
2. Camerini, P., Galbiati, G., Maffioli, F.: Random pseudo-polynomial algorithms for exact matroid problems. J. Algorithms **13**, 258–273 (1992)
3. Lovász, L.: Combinatorial Problems and Exercises. North-Holland, Amsterdam (1979)
4. Megiddo, N.: Combinatorial optimization with rational objective functions. Math. Oper. Res. **4**, 414–424 (1979)
5. Mulmuley, K., Vazirani, U., Vazirani, V.: Matching is as easy as matrix inversion. Combinatorica **7**, 105–113 (1987)
6. Naor, J., Shachnai, H., Tamir, T.: Real-time scheduling with a budget. Algorithmica **47**, 343–364 (2007)

7. Papadimitriou, C.H., Yannakakis, M.: The complexity of restricted spanning tree problems. J. ACM **29**, 285–309 (1982)
8. Ravi, R., Goemans, M.X.: The constrained minimum spanning tree problem. In: Karlsson, R., Lingas, A. (eds.) Proceedings of the 5th Scandinavian Workshop on Algorithms and Theory. Lecture Notes in Computer Science, vol. 1097, pp. 66–75. Springer, Berlin (1996)
9. Schrijver, A.: Theory of Linear and Integer Programming. Wiley, New York (1986)
10. Schrijver, A.: Combinatorial Optimization: Polyhedra and Efficiency. Springer, Berlin (2003)

Motifs in Networks

Karsten Weihe

Abstract Motifs are a useful approach to network analysis. Small, local structures are counted and interpreted. This chapter is a short tour round selected domains in which motif analysis has been successfully applied so far, with a focus on work in which the author's research group has been involved.

1 What Are Motifs, Roughly?

Motifs are "patterns of local interconnection" or, alternatively, "the simple building blocks of complex networks."

2 Graphs or Networks?

First a remark on terminology. Technically, we will use the terms *graph* and *network* synonymously. However, whenever we speak of a *network*, we mean something large; something composed of many interconnected entities; something that reflects the structure of something else: social networks, language-based networks, communication networks, traffic networks, and the like. We consider undirected and directed networks.

3 What Are Motifs, Exactly?

For a *motif analysis* of networks, we select a set of *motifs*. A motif is nothing but a connected graph with a specific structure, so two selected motifs should not be

---

K. Weihe (✉)
Fachbereich Informatik, Technische Universität Darmstadt, Hochschulstraße 10,
64289 Darmstadt, Germany
e-mail: weihe@cs.tu-darmstadt.de

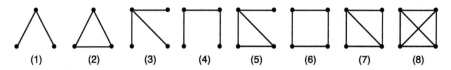

Fig. 1 The eight undirected motifs on three and four nodes, respectively

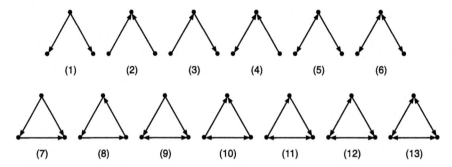

Fig. 2 The directed motifs on three nodes. A *double arrow* stands for two opposite arcs.

isomorphic. Usually, but not exclusively, motifs in directed networks are directed, and motifs in undirected networks are undirected.

Typically, motifs are small, not to say tiny. In many cases, the graphs on three and four nodes are the selected motifs. See Figs. 1 and 2. However, the general concept is not limited to small motifs, not even to a finite set of motifs.

The *motif signature* of a network counts the number of occurrences (a.k.a. matches) of each selected motif in that network. More specifically, an *occurrence* of a motif is a set of nodes in the network such that the subgraph induced by this node set is isomorphic to that motif. As no two selected motifs are isomorphic, no set of nodes can be an occurrence of more than one selected motif.

To be counted simultaneously, two occurrences of a motif may or may not be allowed to share nodes and, if so, may or may not be allowed to share edges. However, determining the maximum number of node-disjoint or edge-disjoint motifs is $\mathcal{NP}$-hard even for very small motifs (look out for the inherent independent set problem!). In contrast, a flat enumeration of all occurrences is straightforward, at least for sufficiently small motifs (for larger motifs, statistical estimation strategies have been developed). Therefore, disjointness conditions are not too popular, and throughout this chapter, we will not consider them, either. Nevertheless, as we will see, even flat enumeration yields interesting, useful results.

It should be mentioned that motifs are often defined slightly differently, yet equivalently. More specifically, the motif signature of a real network from the application is compared with the motif signature of a *null model*, that is, the average over all random graphs that share some selected essential basic characteristics with the real networks from that application. In this alternative definition, *motifs* and *anti-motifs* are those selected subgraphs that occur in the real networks significantly more/less often than in the null model. Typically, the average of all random networks is not computed exactly but approximated by randomly generated networks. These random networks are typically not generated from scratch. Instead, the examined real-world network is taken and modified by repeated small, random steps that maintain the essential basis characteristics. For example, the *single-node characteristics*, which comprise the in-degrees and the out-degrees of all nodes, is quite easy to maintain by simple, obvious, random local modifications and often taken as an essential basic characteristic.

Overall, quite a simple concept, isn't it? But surprisingly useful, as we will see in the sections to follow.

4 Biology

Originally, motif analysis was developed in the realm of computational molecular biology, in particular, *self-regulatory networks*, where the actors (genes, neurons, chemical substances, etc.) affect each other's functions.

In various types of self-regulatory networks, two motifs occur drastically more often than in the respective null model, by far more than the other motifs: the *feed-forward loop* (#7 in Fig. 2) and the *bi-fan* (#1 in Fig. 3).

Our first concrete example is a well-investigated case of self-regulatory networks, called *gene(tic) regulatory networks*. The nodes are the genes of an individual from any species. A directed arc $x \to y$ means that the product of x interacts with the promoting part of y. Natural neural networks is another class where these two motifs are outstanding.

In the following, we will pick out the feed-forward loop and its functionality in gene regulatory networks. Each of the three arcs either expresses a positive (*activating*) or a negative (*repressing*) impact. Let x, y, and z denote the nodes such that $x \to y$, $y \to z$, and $x \to z$ are the arcs.

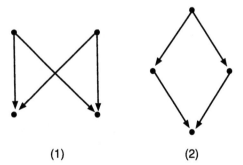

Fig. 3 The bi-fan and the bi-parallel motifs

In biology, two types of feed-forward loops are distinguished, *coherent* and *incoherent* ones, and either type has a specific function. An occurrence of the feed-forward loop is *coherent* in two cases: (1) $x \to z$ is positive and the other two arcs are of the same type, that is, both positive or both negative; (2) $x \to z$ is negative and the other two arcs are of different types. Put in more abstract terms, coherent means that the sign of $x \to z$ equals the product of the signs of $x \to y$ and $y \to z$. In some occurrences of the feed-forward loop, the two signals arriving at z are processed in an AND logic, in the other cases, in an OR logic. In total, we obtain 16 different configurations, and eight of them are coherent. Each of the 16 configurations realizes a Boolean formula:

$x \to y$	$y \to z$	$x \to z$	AND	OR	Coherent
+	+	+	$x \wedge y$	x	Yes
+	+	−	0	y	No
+	−	+	$x \wedge \bar{y}$	$x \vee \bar{y}$	No
+	−	−	$\bar{x}$	$\bar{x} \wedge \bar{y}$	Yes
−	+	+	x	$x \vee y$	No
−	+	−	$\bar{x} \wedge y$	$\bar{x}$	Yes
−	−	+	x	$x \vee \bar{y}$	Yes
−	−	−	$\bar{x} \wedge \bar{y}$	$\bar{x} \vee \bar{y}$	No

This table just scratches the surface. To understand the real functionality of the feed-forward loop, an analysis of the chemical kinetics was necessary, which is far beyond the scope of this chapter. For conciseness, we briefly touch upon but one important, representative effect: controlled delay and acceleration. If y were not present, a change of the Boolean state of x would result in a change of the Boolean state of z within a certain delay time. Now, each type of *coherent* feed-forward loop has a specific delay effect, both in the AND and the OR case. In four out of the eight possible cases, the delay applies whenever x switches from false to true, in the other four cases, whenever x switches from true to false. In contrast, certain types of *incoherent* feed-forward loops accelerate the process.

Motif analysis has been applied in various subdisciplines of biology. The next, short, example is from ecology. In a *food web*, the nodes are species. A directed arc points from a predator to her prey. Here, the *three-chain* (#3 in Fig. 2) and the *bi-parallel motif* (#2 in Fig. 3) are outstanding. The large frequency of the bi-parallel motif simply means that, by tendency, species that are eaten by common species, also feed on common species.

These examples already demonstrate a general aspect: motif analysis has a quantitative and a qualitative side. The number of occurrences of a motif may say something important about the network as a whole (*quantitative* side), and the outstanding motifs have specific functions or reveal insightful phenomena (*qualitative* side). The feed-forward loop example also demonstrates how motif analysis may put a spotlight on network structures and how this may inspire deeper research to understand the subtle and manifold contributions of local structures to the system as a whole.

5 Co-Authorship Networks

We first discovered the power of motif analysis in our work on co-authorship networks. The nodes represent the authors, and two authors are connected by an undirected edge if they co-author at least one joint publication. So, each publication induces a clique on its authors.

We considered the motifs in Fig. 1. For each publication p, we took the total number $c(p)$ of citations of p and computed the *impact* of p in two versions, $i_1(p)$ and $i_2(p)$: $i_1(p)$ is simply $c(p)$ taken as is, $i_1(p) = c(p)$, and $i_2(p)$ is $c(p)$ divided by the number of authors (-1). Based on that, the *impact* of an edge e is defined in four different ways: $i_{1a}(e)$ and $i_{2a}(e)$ sum up the impacts $i_1(p)$ and $i_2(p)$, respectively, of all publications p that contribute to e; $i_{1b}(e)$ and $i_{2b}(e)$ are just $i_{1a}(e)$ and $i_{2a}(e)$ divided by the number of publications contributing to e. The impact of an occurrence of a motif is the average over the impacts of the edges in this occurrence (again these four versions). To obtain the null model, we did not change the structure of the network but randomized the citation frequencies, because citation frequencies was our focus.

Generally speaking, we found that the local patterns of collaboration reveal effects that cannot be explained by the impacts of the involved authors alone. They cannot be explained, either, by simple structural properties such as node degrees. The most striking result is this: the box motif (#6 in Fig. 1) attracts an unexpected number of citations with respect to all four versions defined above. It is important to note that #6 alone is outstanding in this respect, #7 and #8 are rather inconspicuous.

Of course, we looked at the high-impact occurrences of #6. It turned out that three factors largely explain our result: (1) the high-impact occurrences of the box motif have a suspiciously high probability that the two out of the four authors with highest citation counts (the "seniors") are adjacent; (2) the time duration from the rise of the first edge to the rise of the last edge is particularly high, higher than in any other motif, particularly high in the heavy-impact occurrences; (3) the *betweenness values* of the nodes are noticeably high, many high-impact occurrences of the box motif serve as bridges in the network.

Is there any sociological explanation for these findings? In other words, is the box motif a functional block of co-authorship networks like the functional blocks in biological networks? Most probably, the answer is yes. However, the data on which we did our study do not contain the information needed to continue research from a sociological perspective—and, to continue this thread of research in the future, we first have to find interested sociologists (expressions of interest are welcome).

6 Comparing Networks via Motifs

Milo et al. compared directed networks "from biochemistry, neurobiology, ecology, and engineering;" specifically, transcriptional gene regulation networks, ecological food webs, natural neural networks, electronic circuits, and snapshots from the world wide web. In each case, they analyzed a number of networks and compared the motif signature of each network with the motif signatures of random networks (with the same single-node characteristics, cf. Sect. 3).

The overall result is this: like a fingerprint of the domain, the networks from that domain exhibit domain-specific deviations from the motif signature of the null model. So, the deviations are rather similar for networks from a common domain, and rather different for networks from different domains.

An interesting detail is that the motif signatures of gene regulation networks and natural neural networks share remarkable details. The authors suspect that this is not coincidental. For example, the large number of occurrences of the feed-forward loop (#7 in Fig. 2) in both types of networks leads the authors to the hypothesis that this motif "may play a functional role in information processing" in networks of various types.

Another interesting detail is that electronic circuits with fundamentally different function seem to be distinguishable by their motif signatures. The authors try to interpret their statistical results in view of the functions of the individual circuits. However, they admit this cannot be more than a mere speculation; further studies would be necessary for a clearer view.

7 Co-Occurrence Networks of Texts

In Sect. 6, we discussed the observation that classes of networks may be distinguished from each other by their motif signatures. This insight was our motivation, and co-occurrence networks our vehicle. However, we do not compare networks with null models but directly with each other.

A *co-occurrence network* is induced by an unstructured prose text of any kind, written in any natural language. The nodes represent the words that occur in the text and are regarded as relevant. For example, for some research questions it is reasonable to regard ubiquitous words such as conjunctions as irrelevant. On the other hand, for statistical purposes like motif analysis, highly unusual words might be irrelevant. An edge indicates that the two connected words appear close to each other somewhere in the text. The exact definition of "close to each other" is a degree of freedom. For example, it may mean "in the same sentence" or "not more than k words in between." More generally, it may mean that the two words occur "close to each other" not just once but significantly often in the text.

For our study, we focused on co-occurrence within at least one sentence. The relative order of the words in a sentence does not matter, so the edges in our model are undirected. We incorporated a core language vocabulary comprising the 5,000 most frequent words of the language. We found good evidence that, qualitatively, our results are stable over a range from 1,000 up to 20,000 words, as long as indeed the most frequent words of the language are chosen.

We took real text documents, written by humans. From each real text, we generated artificial texts as sequences of random sentences. For that, we used the *n-gram* concept. This concept may be applied on the level of single characters or on the level of single words. Such an application constructs random words and sentences, respectively. Therefore, we applied it on the word level. We inserted an additional, auxiliary word in the vocabulary, which separates two successive sentences. Hence, the end of a sentence needs no special treatment, it is just another word. A Markov chain of order n constructs the artificial text word for word. Roughly speaking, the original text T induces a conditional probability for each word w_n given that $w_1, w_2, \ldots, w_{n-1}$ were the words chosen last. Based on these probabilities, the Markov chain chooses the next word in each step. We generated n-gram texts for $n = 2, 3, 4$.

The real texts were written in the English, German, French, Indonesian, Farsi, and Lithuanian languages. In each case, we found that the frequency of the clique motif (#8 in Fig. 1) suffices to clearly distinguish between original and artificial text. In other words, for an unknown text, the motif analysis suffices to determine whether the text is human-written or n-gram generated.

We compared our results from motif analysis with the global network parameter *transitivity*. For undirected networks, transitivity is defined as the number of closed triangles divided by the number of node triples with at least two edges (the number of occurrences of #2 in Fig. 1 divided by the number of occurrences of #1 and #2 together). Transitivity turned out to be suitable for distinction as well, however, the effect is not so strong.

Moreover, motifs allow a *qualitative* analysis. The chain motif and the box motif (#4 and #6 in Fig. 1) occur substantially more frequently in the original texts than in the artificial texts. This can largely be explained by linguistic phenomena:

- The high frequency of the chain motif in the human-written texts reflects *polysemy*, that is, one word may have different meanings in different contexts. For example, we found the occurrence *Democrats—Social—Sciences—Arts*, in which *Social* is polysemous. Likewise, *One* is polysemous in the chain *Number—One—Formula—Championship*. An example where a word has two completely different (yet related) meanings is the chain *Abraham—Lincoln—Nebraska—Iowa*.
- Analogously, the box motif reflects synonymy, that is, two different words have (nearly) the same meaning. Two typical examples might explain the relation to the box motif without further commentary: *winning—award—won—price* and *wrote—article—published—poems*.

We are going to continue this thread of research in the guiding theme C1, *motif analysis of text-based graphs*, of the DFG research training program AIPHES (www.aiphes.tu-darmstadt.de).

8 Peer-to-Peer Networks

Together with researchers from the telecommunication camp, Krumov, then a Ph.D. student of mine, applied the concept of motifs to peer-to-peer networks.

The nodes of the network are mobile devices, hot spots, etc., the edges are communication channels. It is assumed that the node set of the network does not change (*steady state*). The focus of this research is on *live streaming networks*. The same content, for example a video, is sent (*streamed*) from one sender to many recipients. Due to a strictly limited bandwidth, it is not sent to all of them via direct connections. Some recipients are indeed directly connected with the sender and receive the content via their direct connections. They distribute the content further via direct connections to other recipients, the latter ones distribute the content even further to yet other recipients, and so on. So far, this amounts to one directed tree, which is rooted at the sender. However, the content can be split into several partial streams (*stripes*). Therefore, the big picture consists of several directed trees, all rooted at the sender. Each direct connection between two nodes may be used by one or more trees, even in opposite directions.

The optimization problem is to reorganize the direct connections such that the *resilience* against failures and hostile attacks is maximized. Resilience is quite a fuzzy concept and may be formally defined in different ways. In the discussed work, the definition is based on a worst-case scenario: to which extent can the entire streaming process be damaged if an adversary chooses k, say, nodes and removes them from the network. It is not too hard to see that the optimal topology for a single stripe is a perfectly balanced tree such that each internal node has the maximal possible out-degree. The sum of the out-degrees of a node in all stripes is restricted by the node's bandwidth.

The directed motifs on three nodes were considered. Of course, in trees, only the first and the third motif in Fig. 2 occur at all. The available bandwidth induces an "optimal" mixture of these two motifs. This motif signature then translates into *local balance operations*. Figure 4 shows two exemplary operations.

On each node, a local process is running, which analyzes the local connection structure around the node and tries to enforce its optimal local motif signature in competition with the processes on the other nodes. Another process on each node answers requests of other nodes in the vicinity regarding the current local structure.

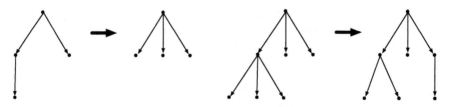

Fig. 4 Two examples of local transformations in peer-to-peer networks to drive the local structure around each peer towards the optimal motif signature

In a steady state, the shapes of the resulting trees converge to a remarkably good resilience—competitive, if not superior, to other approaches. The management overhead on each node is negligible. The true step forward is this: the motif approach is purely local; no central steering is necessary; no node needs any knowledge about the network beyond its immediate vicinity. Since it is not necessary, a node indeed does not receive any piece of non-local information from its neighbored nodes. Consequently, no node has the information that is necessary to harm the network deliberately and successfully.

9 Labeled Motifs

So far, two different occurrences of the same motif within the same network are indistinguishable. However, they do not necessarily have the same function. In fact, two different functions may, coincidentally, induce the same local structure. In these cases, the nodes and/or the edges are assigned labels to indicate that they are of different types.

A concrete example: the nodes are the words found in a given text, and the edges indicate semantic relationships. The nodes are labeled by their word type: noun, verb, adjective, etc.; the edges are labeled according to the type of relation. For instance, a noun may be related to a verb as a subject or as an object or as the main word of an adverbial phrase.

10 Perspectives

Even this small selection of applications should have given a good impression of this generic methodology and of its perspectives for network analysis and even for constructive network problems in a broad variety of domains. We expect many further useful and insightful results from old and new application domains in the future.

11 Literature

The quotes in the first sentence of Sect. 1 are taken from [7] and [8], respectively.

Motif analysis of biological networks is now a broad, mature field. Each of [1, 8, 9] gives a good first impression. Section 4 is based on [7]. The brief discussion of the function of the feed-forward loop is based on [5, 6]. There, further functions of this motif within gene regulatory networks are discussed. The concepts as well as the literal quotes in Sect. 6 are taken from [7].

See [2–4] for further details of the work presented in Sects. 5, 7, and 8, respectively.

References

1. Alon, U.: Network motifs: Theory and experimental approaches. Nat. Rev. Genet. **8**, 450–461 (2007)
2. Biemann, C., Roos, S., Weihe, K.: Quantifying semantics using complex network analysis. In: Proceedings of the 24th International Conference Computational Linguistics (COLING 2012)
3. Krumov, L., Andreeva, A., Strufe, T.: Resilient peer-to-peer live-streaming using motifs. In: 11th International Symposium on a World of Wireless Mobile and Multimedia Networks (WoWMoM 2010)
4. Krumov, L., Fretter, C., Müller-Hannemann, M., Weihe, K., Hütt, M.-T.: Motifs in co-authorship networks and their relation to the impact of scientific publications. Eur. Phys. J. B **84**, 535–540 (2011)
5. Lipshtat, A., Purushothaman, S.P., Iyengar, R., Maáyan, A.: Functions of bifans in context of multiple regulatory motifs in signaling networks. Biophys. J. **94**(7), 2566–2579 (2008)
6. Mangan, S.: Structure and function of the feed-forward loop network motif. Proc. Natl. Acad. Sci. USA **100**(21), 11980–11985 (2003)
7. Milo, R., Shen-Orr, S., Itzkovitz, S., Kashtan, N., Chklovskii, D., Alon, U.: Network motifs: Simple building blocks of complex networks. Science **298**, 824–827 (2002)
8. Schreiber, F., Schwöbbermeyer, H.: Motifs in biological networks. In: Stumpf, M.P.H., Wiuf, C. (eds.) Statistical and Evolutionary Analysis of Biological Networks, pp. 45–64. Imperial College Press, London (2010)
9. Wong, E.A., Baur, B., Quader, S., Huang, C.-H.: Biological network motif detection: Principles and practice. Brief. Bioinform. **13**(2), 202–215 (2011). doi:10.1093/bib/bbr033

Graph Fill-In, Elimination Ordering, Nested Dissection and Contraction Hierarchies

Ben Strasser and Dorothea Wagner

Abstract Graph fill-in, elimination ordering, separators, nested dissection orders and tree-width are only some examples of classical graph concepts that are related in manifold ways. This essay shows how contraction hierarchies, a successful approach to speed up Dijkstra's algorithm for shortest paths, fits into this series of graph concepts. A theoretical consequence of this insight is a guarantee for the size of the search space required by Dijkstra's algorithm combined with contraction hierarchies. On the other hand, the use of nested dissection leads to a very practicable variant of contraction hierarchies that can be applied in scenarios where edge lengths often change.

1 Introduction

We begin by reviewing some relevant basic graph-theoretic concepts before we discuss contraction hierarchies in detail.

1.1 Graph Fill-In and Elimination Ordering

Let $G = (V, E)$ be a connected, undirected simple graph with n vertices and m edges. A graph is *chordal* if it does not contain a chordless cycle of length at least four. Chordal graphs can be characterized by the concept of a perfect elimination ordering. A vertex v is called *simplicial* in G if the neighbors of v form a clique in G. It is known that every chordal graph contains a simplicial vertex and that removing (or eliminating) a simplicial vertex and its incident edges from a chordal graph yields

---

B. Strasser · D. Wagner (✉)
Institut für Theoretische Informatik, Karlsruher Institut für Technologie, Am Fasanengarten 5, 76131 Karlsruhe, Germany
e-mail: dorothea.wagner@kit.edu

B. Strasser
e-mail: strasser@kit.edu

© Springer International Publishing Switzerland 2015
A.S. Schulz et al. (eds.), *Gems of Combinatorial Optimization and Graph Algorithms*, DOI 10.1007/978-3-319-24971-1_7

a chordal graph. A *perfect elimination ordering* of G is a vertex ordering r, i.e., a bijective function $r : V \to \{1, \ldots, n\}$ where each vertex v is simplicial in the graph induced by the set of vertices $u \in V$ with $r(u) \geq r(v)$. Accordingly, chordal graphs can be characterized as those graphs that have a perfect elimination ordering.

The *elimination game* on G considers a vertex ordering r and eliminates the vertices and all incident edges according to r. When a vertex v is eliminated, all neighbors of v are connected by additional edges to form a clique. The ordering r is called an *elimination ordering*. Let F denote the set of those additional edges. The *filled* graph G^+ is the supergraph of G with edge set $E \cup F$. Obviously, G^+ is a chordal supergraph of G, and the ordering r is a perfect elimination ordering of G^+. Each vertex v together with its neighbors u with $r(u) > r(v)$ form a maximal clique in G^+. The related minimization problem of finding a vertex ordering r such that the size of F is minimum is known as the *minimum fill in* or the *chordal graph completion problem*. Figure 1a depicts a graph G with vertices labeled according to r, the corresponding chordal supergraph G^+ and the induced cliques. The *elimination tree* T for an elimination ordering r of G is a rooted tree with vertex set V and defined by assigning to each vertex v as parent its neighbor u in G^+ with $r(u) > r(v)$ and $r(u)$ minimum. The unique vertex v, which does not posses such a neighbor, is the root of T. The *tree-depth* $\text{td}(G)$ of G is defined as the minimum elimination tree height over all elimination orderings r. Figure 1b depicts the elimination tree related to the elimination order in Fig. 1a.

Chordal graphs are tightly coupled with the concepts of *tree-decomposition* and *tree-width*. Indeed it is common to characterize the tree-width $\text{tw}(G)$ of a graph G using its chordal supergraphs. That is, $\text{tw}(G)$ is the largest number such that, for all chordal supergraphs G^+ of G, $\text{tw}(G) < k$ holds, where k is the maximum clique size of G^+. A common technique to compute some tree-decomposition of G consists in choosing a vertex ordering r of G and constructing the chordal supergraph G^+ induced by the elimination game on G with ordering r. Then the maximal cliques

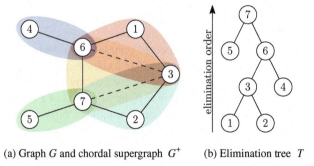

(a) Graph G and chordal supergraph G^+ (b) Elimination tree T

Fig. 1 The left figure depicts the graph G and its chordal supergraph G^+. The right figure depicts the corresponding elimination tree. The vertices are numbered according to r. In the left figure the solid lines denote the edges of G. When constructing the corresponding chordal supergraph G^+ the dashed edges are added. The colored regions are the maximal cliques in the chordal graph G^+ and the respective vertex sets of a tree decomposition of G.

of G^+ form the sets of a tree-decomposition of G. The size of the maximum set in this tree-decomposition is called its width. So finding a tree-decomposition of G with minimum width $\text{tw}(G)$ is equivalent to finding a chordal supergraph G^+ with smallest maximum clique size.

1.2 Nested Dissection

A *balanced separator* S of size $|S|$ is a subset of V that induces a partition V_1, V_2 of $V \setminus S$ such that $|V_1| \leq \alpha n$, $|V_2| \leq \alpha n$, with $\frac{1}{2} \leq \alpha < 1$ and there exists no edge $\{x, y\} \in E$ with $x \in V_1$ and $y \in V_2$. The subgraphs G_1 and G_2 induced by V_1 and V_2 are the sides of the separator. The objective is to compute a balanced separator of minimum or at least bounded size, i.e., $|S| \in O(n^\beta)$ where $0 \leq \beta < 1$.

A common technique to compute an elimination ordering of a graph G is *nested dissection*. The idea is simple: Recursively partition the graph using small balanced separators. Then consider the elimination ordering of the vertices induced by the recursive structure of the partition where each of the two sides of the separator is eliminated first, and then the separator vertices are eliminated. More precisely, the nested dissection starts by determining a small balanced separator S of G. Then recursively elimination orderings r_1 and r_2 are determined for the two sides of S. The elimination ordering r of G consists of concatenating r_1, followed by r_2, followed by S. The base case of the recursion is reached when the graph has less than some constant number of vertices. In this case the vertices are taken in some arbitrary ordering. Nested dissection is illustrated in Fig. 2. It is known that every graph G with recursive $O(n^\beta)$ balanced separators has a tree-depth in $O(n^\beta)$. Using a nested dissection ordering yields a corresponding elimination tree.

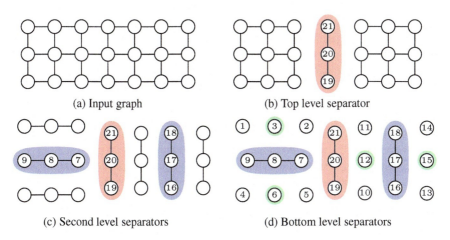

Fig. 2 Steps of the nested dissection procedure. Vertices are labeled according to the induced elimination ordering.

There is a classical approach to use chordal graph completion and elimination orderings to speed up the *Gaussian elimination* of sparse symmetric matrices $M \in \mathbb{R}^{n \times n}$. Given M, one can construct a graph $G = (V, E)$ with $V = \{v_1, \ldots, v_n\}$ and $\{v_i, v_j\} \in E$ if and only if $M_{ij} \neq 0$. Eliminating the ith row and ith column in M corresponds to the elimination of v_i in G. Eliminating the rows and columns according to a good elimination ordering of G assures that many zero-entries are conserved in M during the execution of the Gaussian algorithm. This can be exploited to significantly improve the running times.

1.3 Two-Phase and Three-Phase Shortest Path Computation

In a directed graph $G = (V, A)$ with an arc weight function $w : A \to \mathbb{R}_{\geq 0} \cup \{\infty\}$, an st-path P is defined as a sequence of vertices $v_1 \ldots v_k$, such that $s = v_1, t = v_k$, $(v_i, v_{i+1}) \in A$, and its length is defined as $w(P) := \sum_{i=1}^{k-1} w(v_i, v_{i+1})$. The input of the *classic shortest path problem* consists of a directed graph G, an arc weight function w, and two vertices s and t. The output is an st-path of minimum length. This problem can be solved using Dijkstra's algorithm in near-linear running time. However, for graphs with millions of vertices and edges this is not fast enough. To accelerate the shortest path computation for graphs that do not change often, one can apply a *two-phase shortest path computation*: In a first *preprocessing phase* only G and w are known and some auxiliary data are computed. In a second phase, the *query phase*, s and t become available and a shortest st-path is computed. The second phase can use the auxiliary data. For example, one may apply a variant of Dijkstra's algorithm that makes use of the previously determined auxiliary data to reduce its search space. Note that the auxiliary data are independent of s and t and can therefore be used again and again for many queries. However, a central assumption of this approach is that G and w do not change between queries.

A prominent application of the two-phase shortest path computation is route planning in transportation networks where many queries need to be answered quickly. The key observation is that it is enough to make the query phase fast, while the preprocessing phase can be slow. However, the assumption that G and w do not change between queries is not always true. In road graphs, G is indeed mostly constant but w may change due to the current traffic situation or individual restrictions by the users. The corresponding shortest path problem is called *customizable route planning*. For such scenarios it is more adequate to apply a *three-phase shortest path computation*. Here the preprocessing phase is split into two phases: A weight-independent preprocessing phase that computes weight-independent auxiliary data based only on G, and a *customization phase* where the auxiliary data are augmented depending on w. The weight-independent preprocessing phase may still be slow, but the customization phase should be reasonably fast.

2 Contraction Hierarchies

In order to simplify the presentation, only undirected graphs are considered in the following sections. Moreover, for technical reasons we assume that for each pair of vertices s, t the shortest st-path is unique. However, we want to point out that all these results are developed and presented for directed graphs in the original papers and are applied in shortest path computations and its applications in route planning as stated in Sect. 1.3.

An ingredient of many shortest path acceleration techniques are *shortcuts*, i.e., additional edges computed in the preprocessing phase to build the auxiliary data or at least a part of it. The idea consists of selecting an important path $v_1 \ldots v_k$ and adding an additional edge to the graph from v_1 directly to v_k with the weight $\sum_{i=2}^{k} w(v_{i-1}, v_i)$. Acceleration techniques use shortcuts to prevent Dijkstra's algorithm from visiting all intermediate vertices. The speed-up achieved depends on the choice of paths bypassed by shortcuts, and it is crucial that the number of shortcuts is small. *Contraction Hierarchies (CH)* is an elegant and effective two-phase shortest path computation based primarily on shortcuts.

2.1 Two-Phase Contraction Hierarchies

The two phases of CH consist of a systematic preprocessing approach to add shortcuts and a sophisticated way to perform the query phase. The name giving operation *contraction* selects a vertex x from G, removes x from G and adds, if necessary, the edge $\{y, z\}$ if and only if yxz is a shortest yz-path in the current graph with respect to w. The edge $\{y, z\}$ is assigned the weight $w(y, z) = w(y, x) + w(x, z)$. In the CH preprocessing phase, contraction is applied iteratively according to some previously chosen *contraction ordering* $r : V \to \{1, \ldots, n\}$ of the vertices in G. Note that this process bears some similarity to the elimination game introduced in Sect. 1.1. However, not all neighbors of the contracted vertex v are connected by a new edge like in the elimination game. In the following, we denote the set of all original edges and added shortcuts by E^+, and let $G^+ = (V, E^+)$. See Fig. 3. Commonly, the graph G^+ together with the corresponding edge weight w is called a *contraction hierarchy*. For achieving a significant speed-up the choice of the contraction ordering is crucial.

The CH query phase works similar to the bidirectional variant of Dijkstra's algorithm. The shortcut augmented graph $G^+ = (V, E^+)$ together with r induces a directed graph $G^\uparrow = (V, E^\uparrow)$, where $E^\uparrow$ contains all edges (y, x) of E^+ such that y comes before x in the contraction ordering. The subgraph of $E^\uparrow$ reachable from a vertex x is called the *upward search space* of x. To determine a shortest st-path a bidirectional variant of Dijkstra's algorithm is run. The forward search is restricted to the search space of s, whereas the backward search is restricted to the search space of t. As G is assumed to be connected, these two search spaces overlap. See Fig. 4.

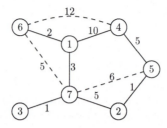

Fig. 3 Graph and CH shortcuts. The vertices are numbered according to the contraction ordering. The *solid lines* are edges in G and the *dashed lines* are inserted shortcuts. Edges are annotated with their weights. When contracting vertex 1 shortcuts from 6 to 4 and from 6 to 7 are added. However, no shortcut from 4 to 7 is added because the path $4 \to 1 \to 7$ has length 13, which is longer than the path $4 \to 5 \to 2 \to 7$ with length 11. For the same reason no edge is added between vertices 5 and 6 when contracting vertex 4.

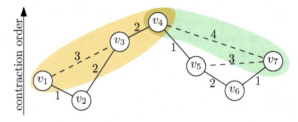

Fig. 4 An illustrated CH query. G is the line graph with vertices $v_1 \ldots v_7$. The vertical position of a vertex corresponds to its position in the contraction order. The *solid lines* represent the edges in G with their weights. The *dashed lines* are the added shortcuts in G^+. The *orange area* is the search space of v_1 and the *green area* is the search space of v_7. Both search spaces meet in v_4. The CH finds the path v_1, v_3, v_4, v_7 of length 9 in G^+, which has the same length as the shortest path $v_1 \ldots v_7$ in G.

Lemma 1 *The CH query phase is correct, i.e., it computes a shortest st-path.*

Proof The CH query induces an st-path $su_1 \ldots u_k q d_1 \ldots d_l t$ in G^+ such that the vertices $s, u_1 \ldots u_k, q$ appear in increasing contraction ordering, i.e., $r(s) < r(u_1) < \cdots < r(u_k) < r(q)$, and the vertices $q, d_1 \ldots d_l, t$ appear in decreasing contraction ordering, i.e., $r(q) > r(d_1) > \cdots > r(d_l) > r(t)$. Such an st-path is called *up-down st*-path. Note that one of the two subpaths might be empty or even both if $s = t$.

It remains to prove that for each pair of vertices s and t there exists a shortest up-down st-path in G^+. Consider an arbitrary shortest path $P = v_1 \ldots v_h$ in G and iteratively construct an up-down path of equal length in G^+ as follows. Either P already is an up-down path in G^+, or there exists a vertex v_i in P such that $r(v_{i-1}) > r(v_i)$ and $r(v_i) < r(v_{i+1})$. That is, in the CH preprocessing phase v_i was contracted before v_{i-1} and v_{i+1}. Further, $v_{i-1} v_i v_{i+1}$ is a shortest path as it is a subpath of P. Therefore there exists a shortcut from v_{i-1} to v_{i+1} in G^+. Remove v_i from the path, insert this shortcut instead and apply this operation iteratively for the resulting path. As the number of vertices in the path shrinks in each iteration, it is guaranteed that the construction eventually ends with an up-down path that is also a shortest path. □

Remarkably, the correctness of the CH query is still guaranteed if additional "unnecessary" edges are added during the preprocessing phase, as long as the weight of such an edge (y, z) is assigned the length of a shortest yz-path.

2.2 Implementing Contraction Hierarchies

Determining whether yxz is a shortest yz-path is called *witness search*. The straightforward approach consists of running Dijkstra's algorithm on $G\setminus\{x\}$ to find a shortest yz-path in $G\setminus\{x\}$. However, this can be too slow to work on huge graphs. One idea is to abort the witness search at some point and insert the shortcut (y, z) anyway, independently of whether yxz is a shortest yz-path or not. This might result in adding unnecessary shortcuts which make the query phase slower, but fortunately not incorrect.

A heuristic approach to come up with a "good" contraction hierarchy consists in ordering the vertices by ascending "importance." Unfortunately, there is no universal definition for importance of a vertex in a graph. In road graphs "importance" is usually based on the intuition that a vertex on a highway is important whereas a dead-end of a street in a rural area is unimportant.

A contraction ordering may be computed by exploring all possible vertex contractions and greedily picking the vertex that results in the fewest shortcuts and assigning it a low "importance," i.e., we iteratively try to identify dead-end-like structures. This strategy can be refined with further heuristics that try to assure that G is contracted uniformly. A different top-down heuristic consists in iteratively assigning a high "importance" to a vertex that covers many shortest paths, i.e., vertex x that covers as many paths as possible gets highest importance and next the vertex y that covers as many paths as possible not already covered by x is determined, and so on. The idea is that many shortest paths traverse highways. For references, see Sect. 5. Note that these strategies and the quality of the resulting CH depend on the weight function of G. An ordering that leads to a "good" contraction hierarchy for a weight function w_1 can lead to a huge number of shortcuts when used with another weight function w_2 on the same graph G. Even correlated weights such as travel-time and travel-distance in road graphs are not interchangeable in practice.

The next two sections are devoted to these two aspects: How can we determine a contraction ordering for which we can give a guarantee for the size of the search space? How can we construct a weight-independent contraction ordering and design a three-phase shortest path algorithm?

3 Weak Contraction Hierarchies

In the following a generalization of CH is presented for which, depending on the properties of the graphs considered, a guarantee for the size of the search space can be given. Consider a graph G with edge weights w and an arbitrary contraction ordering r. Let $G^+ = (V, E^+)$ denote the contraction hierarchy induced by r with corresponding edge weight w as defined in Sect. 2.1. A graph $H = (V, E_H)$ with $E^+ \subseteq E_H$ and for which the additional edges $\{y, x\}$ are shortcuts is called a *weak contraction hierarchy*. A CH making use of a weak contraction hierarchy is called *weak CH*. Note that a contraction hierarchy determined by CH as originally defined is a *minimum weak contraction hierarchy*, while a *maximum weak contraction hierarchy* contains all possible shortcuts. Moreover, the search space size induced by a maximum weak contraction hierarchy is an upper bound for the search space size induced by any weak contraction hierarchy that is contained therein. In the following, a contraction hierarchy (be it minimum, maximum or any weak contraction hierarchy in between) is denoted by $G^+ = (V, E^+)$.

3.1 A Crucial Observation

The first insight is that the maximum weak contraction hierarchy G^+ for G and w induced by some contraction ordering r, is the chordal supergraph of G with corresponding edge weights w that is obtained by using r as elimination ordering. The next observation is that the vertices in the search space of a vertex x of a weak CH are the ancestors of x in the corresponding elimination tree. Then, instead of applying Dijkstra's algorithm, the query can make use of the elimination tree. Just follow the paths from s and t, respectively, up to the root of the elimination tree. Consider a graph that admits recursive graph separators of size $O(n^\beta)$. Choosing r as a nested dissection ordering yields an elimination tree of depth $O(n^\beta)$. It follows that for every vertex x the number of vertices in the search space of the weak CH using r is in $O(n^\beta)$.

It is well known that planar graphs admit recursive $O(\sqrt{n})$ balanced separators. This induces a weak CH for planar graphs with a search space size guarantee of at most $O(\sqrt{n})$ vertices. The search spaces are not necessarily planar and therefore we can have up to $O((\sqrt{n})^2) = O(n)$ edges in the search space. This results in a running time of $O(n \log n)$ using Dijkstra's algorithm. However, the alternative query procedure that makes use of the elimination tree yields a better worst-case running time of $O(n)$.

3.2 Notes on Some Practical Implications

The theoretical insights discussed above have various practical implications. Most importantly, the bounds on the size of the search space of weak CH based on a nested dissection ordering are completely independent of the edge weights. This is surprising since the performance of CH induced by a weight-dependent ordering, as originally designed, very much depends on the weights, as already mentioned in Sect. 2.2. It is interesting to have a closer look at CH search spaces based on different, i.e., weight-dependent versus weight-independent orderings on one hand, and on minimum versus maximum contraction hierarchies on the other hand. Weak CH using a nested dissection ordering turns out to be also useful in practice. One reason is that road graphs are somehow close or at least similar to planar graphs. Therefore one can expect road graphs to have recursive balanced separators of small size. Indeed, experiments show that road graphs have small separators. Therefore one can expect that CH or weak CH using nested dissection orderings work well on road graphs. Experiments show that this is indeed the case. At least, experimental evaluations show that maximum weak contraction hierarchies based on nested dissection orderings are small enough to be useful.

While CH search spaces for (at least almost) minimum contraction hierarchies induced by weight-dependent orderings are small in practice, the search space of CH based on minimum contraction hierarchies obtained using nested dissection orderings are larger. Experiments have further shown that the search spaces of weak CH using maximum contraction hierarchies resulting from weight-dependent orderings are too huge to be computable in practice. The experiments clearly show that contraction hierarchies resulting from the weight-dependent orderings (considered so far) differ from (weight-independent) contraction hierarchies obtained by nested dissection. But how these two worlds relate is not yet fully understood.

4 Customizable Contraction Hierarchies

In a departure from the problem setting considered in the previous two sections, we now consider situations in which the arc weight function w may change between queries due to the current traffic situation or individual restrictions by users. Recall from Sect. 1.3 that an adequate approach in such scenarios consists of a three-phase shortest path computation where the preprocessing is split into two phases, a weight-independent preprocessing phase and a customization phase.

It turns out that weight-independent contraction orderings are a very good basis for a three-phase shortest path computation. The key idea is that a maximum weak contraction hierarchy obtained by a weight-independent ordering already contains all necessary shortcuts for each possible weight function. Accordingly, computing a maximum weak contraction hierarchy based on nested dissection in a first preprocessing phase yields a good basis for the subsequent second and third phases. In

Fig. 5 The triangle $\{x, y, z\}$ is a *lower triangle* of the edge $\{y, z\}$, an *intermediate triangle* of the edge $\{x, z\}$ and an *upper triangle* of the edge $\{x, y\}$.

particular, in the customization phase only the weights of the shortcuts need to be computed. Actually, this second phase plays a decisive role for the design of *Customizable Contraction Hierarchies (CCH)* for three-phase shortest path computation.

4.1 Basic Customization

The first phase of the preprocessing phase just considers the graph G without edge weights and returns an elimination ordering r, the corresponding chordal supergraph $G^+ = (V, E^+)$, the induced upward graph $G^\uparrow = (V, E^\uparrow)$, and the corresponding elimination tree T.

Consider a triangle $\{x, y, z\}$ in G^+ such that $r(x) < r(y) < r(z)$, as illustrated in Fig. 5. The tuple $\{x, y, z\}$ is called a *lower triangle* of $\{y, z\}$, an *intermediate triangle* of $\{x, z\}$, and an *upper triangle* of $\{x, y\}$. The *lower triangle inequality* holds when for every lower triangle $\{x, y, z\}$ the inequality $w(y, z) \leq w(y, x) + w(x, z)$ holds. The proof of Lemma 1, which showed the correctness of the query phase, relies on a weakened variant of the lower triangle inequality; hence, if the lower triangle inequality holds, then the queries are correct. Therefore, the goal of the customization phase is to assign weights to shortcuts with respect to a metric w such that the lower triangle inequality is guaranteed. Initially, each edge of G is assigned its weight according to w, and each shortcut is assigned the weight ∞. Then iterate over all y in the order of increasing $r(y)$. For each y, enumerate for all (y, z) in $E^\uparrow$ the lower triangles $\{x, y, z\}$ by exploring the common neighbors of y and z. If $w(y, z) > w(y, x) + w(x, z)$ for a triangle $\{x, y, z\}$, then $w(y, z)$ is set to $w(y, x) + w(x, z)$. This procedure assigns a weight to every edge that guarantees the lower triangle inequality without modifying the topology of the contraction hierarchy G^+. This procedure is called *basic customization*. Note that the lower triangle inequality does not necessarily imply that the weight of a shortcut $\{y, z\}$ equals the length of a shortest yz-path.

4.2 Perfect Customization

If after the basic customization the weight $w(y, z)$ of a shortcut $\{y, z\}$ is larger than the length of a shortest yz-path, this shortcut is not necessary to ensure correctness of the query phase, as subpaths of a shortest up-down path must be shortest paths.

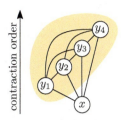

Fig. 6 Upward neighborhood of x. The nodes y_i are all neighbors of x that were contracted after x. By construction, these neighbors must form a clique.

Therefore, the customization phase can be enhanced by a procedure to compute the distances between the endpoints of every edge in G^+ and to subsequently remove all unnecessary shortcuts. This procedure iterates over all x with decreasing $r(x)$ and, for each $\{x, y\}$, enumerates all intermediate and all upper triangles. For each upper and intermediate triangle $\{x, y, z\}$ of $\{x, y\}$ with $w(x, y) > w(x, z) + w(z, y)$, the weight $w(x, y)$ is set to $w(x, z) + w(z, y)$. We will prove that at the end of the iteration, the weight of each edge $\{x, y\}$ corresponds to the length of a shortest xy-path. Such a customization is called *perfect customization*. Although perfect customization costs extra time, it can be beneficial because it results in faster queries as there are less edges in G^+.

Lemma 2 *After perfect customization the weight $w(x, y)$ for each edge $\{x, y\}$ in G^+ equals the length of a shortest xy-path.*

Proof For an inductive proof, consider vertex $x \in V$ and assume that the claim holds for all edges incident to y with $r(y) > r(x)$. Let $y_1, \ldots, y_k$ denote the upper neighbors of x, i.e., $r(y_i) > r(x)$, as depicted in Fig. 6. Because all upper neighbors of x appear after x in the order, we know by induction that the weights of all edges between such neighbors $\{y_i, y_j\}$ (the orange area) equal the length of a shortest $y_i y_j$-path. When an upper or an intermediate triangle of the edge $\{x, y_i\}$ is inspected, either the weight of (x, y_i) is equal to the length of a shortest xy_i-path and there is nothing to prove. Or there exists an xy_i-path P of shorter length, which is an up-down path, because of the basic customization, and contains another neighbor y_j. As x is contracted before all its upper neighbors $y_1, \ldots, y_k$, the vertices $x, y_1, \ldots, y_k$ form a clique in G^+. So there is an edge $\{y_i, y_j\}$, and the length of $xy_i y_j$ is equal to the length of P. Moreover, $\{x, y_i, y_j\}$ is an upper or an intermediate triangle of $\{x, y_i\}$, depending on whether y_i or y_j comes first in the order. □

4.3 Query Phase

The CCH query phase can be performed analogously to the CH query based on Dijkstra's algorithm. Alternatively, one can also make use of the elimination tree. Recall that every vertex in the search space of a query from s to t must be an ancestor

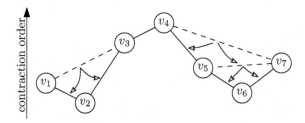

Fig. 7 Path unpacking. The up-down-path $v_1 v_3 v_4 v_7$ in G^+ is unpacked to the path $v_1 \ldots v_7$ in G of equal length. The shortcut $\{v_1, v_3\}$ is replaced by $v_1 v_2 v_3$ by enumerating all lower triangles of the edge $\{v_1, v_3\}$. Similarly $\{v_4, v_7\}$ is unpacked to $v_4 v_5 v_7$. In a subsequent step $\{v_5, v_7\}$ is unpacked to $v_5 v_6 v_7$.

of s or t in the elimination tree. Furthermore, shortcuts are always directed upwards. Therefore, the ancestors of s and t can be enumerated simultaneously, ordered by their position in the contraction ordering, until the root is reached.

The result of the query phase is the distance from s to t and an up-down st-path P in G^+. It remains to clarify how the according shortest path in G can be derived. This can be done by a procedure called *unpacking* that iteratively replaces shortcuts. Consider an edge $\{x, y\}$ on P that is a shortcut. To replace $\{x, y\}$ by a shortest subpath in G enumerate all lower triangles $\{x, y, z\}$ of $\{x, y\}$. By construction, at least one triangle with $w(x, z) + w(z, y) = w(x, y)$ must exist. Therefore the edge $\{x, y\}$ can be replaced by the two subsequent edges $\{x, z\}$ and $\{z, y\}$. Note that $\{x, z\}$ or $\{z, y\}$ may again be shortcuts. However, this unpacking step can be performed recursively. The process is illustrated in Fig. 7.

4.4 Note on Directed Graphs

Remember that we restricted the presentation in the previous sections to undirected graphs in order to simplify notation. However, it turns out that CCH as introduced here can be immediately applied to directed graphs. First notice that a directed graph G with weight function w can be identified with the complete directed graph where arcs not present in G have weight ∞. This graph can be also represented by a complete undirected graph with two weights per edge, a forward and a backward weight. Now, remember that a maximum weak contraction hierarchy G^+ contains all possible shortcuts and that in the first phase, CCH just computes G^+ without weights. In the customization phase, instead of one weight per edge two weights need to be considered, one interpreted as upward weight and the other as downward weight. All procedures can be applied to such a scenario without further modification.

5 Notes on the Literature

Algorithms for route planning in transportation networks have recently undergone a rapid development, leading to methods that are up to several million times faster than Dijkstra's algorithm [13]. Prominent examples besides CH [15, 16] are arc-flags [20], multi-level overlays [19, 24], reach [18], or hub-labels [3, 10]. See also [5] for an overview. Only recently, [6] noticed the connection between contraction hierarchies and the classical graph fill-in problem, which led to a theoretical guarantee for the size of the search space required by weak CH. On the other hand, this insight led to a very practicable variant of contraction hierarchies that can be applied in scenarios where edge lengths often change [11]. See also [12], where some of the theoretical results from [6] are even improved. A different approach to obtain theoretical guarantees for the size of the search space required by speed-up techniques is presented in [1, 2, 4]. The resulting bounds depend on a weight-dependent graph measure called highway dimension.

The basic graph concepts discussed here go back to the 1960s and 70s. It was shown in [14] that every chordal graph contains a simplicial node and that removing (or eliminating) a simplicial node and its incident edges from a chordal graph yields a chordal graph. The connection between chordal graph completion and Gauss elimination was noticed in [22, 23], and the concept of nested dissection goes back to [17] and [21]. Chordal graphs are tightly coupled with the concept of tree-width, and we recommend [7–9] for a survey.

References

1. Abraham, I., Delling, D., Fiat, A., Goldberg, A.V., Werneck, R.F.: VC-dimension and shortest path algorithms. In: Proceedings of the 38th International Colloquium on Automata, Languages, and Programming (ICALP'11). Lecture Notes in Computer Science, vol. 6755, pp. 690–699. Springer (2011)
2. Abraham, I., Delling, D., Fiat, A., Goldberg, A.V., Werneck, R.F.: Highway dimension and provably efficient shortest path algorithms. Technical report MSR-TR-2013-91, Microsoft Research (2013)
3. Abraham, I., Delling, D., Goldberg, A.V., Werneck, R.F.: A hub-based labeling algorithm for shortest paths on road networks. In: Proceedings of the 10th International Symposium on Experimental Algorithms (SEA'11). Lecture Notes in Computer Science, vol. 6630, pp. 230–241. Springer (2011)
4. Abraham, I., Fiat, A., Goldberg, A.V., Werneck, R.F.: Highway dimension, shortest paths, and provably efficient algorithms. In: Proceedings of the 21st Annual ACM–SIAM Symposium on Discrete Algorithms (SODA'10), pp 782–793. SIAM (2010)
5. Bast, H., Delling, D., Goldberg, A.V., Müller–Hannemann, M., Pajor, T., Sanders, P., Wagner, D., Werneck, R.F.: Route planning in transportation networks. Technical report, ArXiv e-prints, (2015). arXiv:1504.05140
6. Bauer, R., Columbus, T., Rutter, I., Wagner, D.: Search-space size in contraction hierarchies. In: Proceedings of the 40th International Colloquium on Automata, Languages, and Programming (ICALP'13). Lecture Notes in Computer Science, vol. 7965, pp. 93–104. Springer (2013)
7. Bodlaender, Hans L.: A tourist guide through treewidth. Acta Cybern. **11**, 1–23 (1993)

8. Bodlaender, Hans L.: Tutorial: A partial k-arboretum of graphs with bounded treewidth. Theor. Comput. Sci. **209**, 1–45 (1998)
9. Bodlaender, H.L.: Treewidth: Structure and algorithms. In: Proceedings of the 14th International Colloquium on Structural Information and Communication Complexity. Lecture Notes in Computer Science, vol. 4474, pp. 11–25. Springer (2007)
10. Cohen, E., Halperin, E., Kaplan, H., Zwick, U.: Reachability and distance queries via 2-hop labels. SIAM J. Comput. **32**(5), 1338–1355 (2003)
11. Dibbelt, J., Strasser, B., Wagner, D.: Customizable contraction hierarchies. In: Proceedings of the 13th International Symposium on Experimental Algorithms (SEA'14). Lecture Notes in Computer Science, vol. 8504, pp. 271–282. Springer (2014)
12. Dibbelt, J., Strasser, B., Wagner, D.: Customizable contraction hierarchies. Technical report, ITI Wagner, Department of Informatics, Karlsruhe Institute of Technology (KIT) (2014). arXiv:1402.0402
13. Dijkstra, E.W.: A note on two problems in connexion with graphs. Numer. Math. **1**, 269–271 (1959)
14. Fulkerson, D.R., Gross, O.A.: Incidence matrices and interval graphs. Pac. J. Math. **15**(3), 835–855 (1965)
15. Geisberger, R., Sanders, P., Schultes, D., Delling, D.: Contraction hierarchies: Faster and simpler hierarchical routing in road networks. In: Proceedings of the 7th Workshop on Experimental Algorithms (WEA'08). Lecture Notes in Computer Science, vol. 5038, pp. 319–333. Springer (2008)
16. Geisberger, R., Sanders, P., Schultes, D., Vetter, C.: Exact routing in large road networks using contraction hierarchies. Transp. Sci. **46**(3), 388–404 (2012)
17. George, A.: Nested dissection of a regular finite element mesh. SIAM J. Numer. Anal. **10**(2), 345–363 (1973)
18. Gutman, R.J.: Reach-based routing: A new approach to shortest path algorithms optimized for road networks. In: Proceedings of the 6th Workshop on Algorithm Engineering and Experiments (ALENEX'04), pp. 100–111. SIAM (2004)
19. Holzer, M., Schulz, F., Wagner, D.: Engineering multilevel overlay graphs for shortest-path queries. ACM J. Exp. Algorithmics **13**(2.5):1–26 (2008)
20. Köhler, E., Möhring, R.H., Schilling, H.: Acceleration of shortest path and constrained shortest path computation. In: Proceedings of the 4th Workshop on Experimental Algorithms (WEA'05). Lecture Notes in Computer Science, vol. 3503, pp. 126–138. Springer (2005)
21. Lipton, R.J., Rose, D.J., Tarjan, R.: Generalized nested dissection. SIAM J. Numer. Anal. **16**(2), 346–358 (1979)
22. Parter, S.V.: The use of linear graphs in Gauss elimination. SIAM Rev. **3**(2), 119–130 (1961)
23. Rose, D.J.: Triangulated graphs and the elimination process. J. Math. Anal. Appl. **32**(3), 597–609 (1970)
24. Schulz, F., Wagner, D., Zaroliagis, C.: Using multi-level graphs for timetable information in railway systems. In: Proceedings of the 4th Workshop on Algorithm Engineering and Experiments (ALENEX'02). Lecture Notes in Computer Science, vol. 2409, pp. 43–59. Springer, (2002)

Shortest Path to Mechanism Design

Rudolf Müller and Marc Uetz

Abstract Mechanism design is concerned with computing desired outcomes in situations where data is distributed among selfish agents. We discuss some of the most fundamental questions in the design of mechanisms, and derive simple answers by interpreting the problem in graph-theoretic terms. Specifically, much of mechanism design is thereby reformulated as shortest path problems.

1 Terminology in Mechanism Design

The basic premise in the design of mechanisms is the presence of a set of noncooperative agents, each of which is equipped with a piece of private information. This private information is called the agent's *type*. The set of all types, one for each agent, effectively defines the input of some optimization problem. There is a principal, henceforth called the *mechanism designer*, in charge of selecting or computing a solution for the optimization problem, referred to as the *outcome*. Each agent has a certain *valuation* for the chosen outcome, which in general depends on her type. The mechanism designer aims to choose an outcome so as to optimize some global objective function, which typically, but not necessarily, depends on the agents' types. A classical example for such a global objective is the so-called utilitarian maximizer, which is the outcome that maximizes the total valuations of all agents. However, the fundamental problem in choosing that outcome is that the mechanism designer does not know the types of the agents. In other words, the (true) input for the problem to be solved is not known a priori. In such situations, the mechanism designer needs to create incentives so that the agents share their private information.

R. Müller
Department of Quantitative Economics, Maastricht University, P.O. Box 616,
6200MD Maastricht, The Netherlands
e-mail: r.muller@maastrichtuniversity.nl

M. Uetz (✉)
Department of Applied Mathematics, University of Twente, P.O. Box 217,
7500AE Enschede, The Netherlands
e-mail: m.uetz@utwente.nl

The best known example of a mechanism is probably *Vickrey's second price auction*: When auctioning a single, indivisible item, it gives the item to the bidder with the highest bid and sets the price for the winner to the second highest bid. This makes truthful bidding a dominant strategy. Like in Vickrey's auction, mechanism design generally allows for payments, also called transfers. Central questions are: Is there a payment rule that incentivizes agents to tell the truth? And which flexibility is there in choosing such payment? Indeed, if payments are taxes, we might want to reduce tax load and make them as small as possible, while in auction settings we might want to increase the revenue for the seller, hence maximize payments. We will answer these questions using basic graph-theoretic concepts.

1.1 Types, Outcomes and Valuations

To fix some notation, we denote by $\{1, \ldots, n\}$ the set of agents and let A be the set of possible outcomes. Outcome space A is allowed to have infinitely many, even uncountably many, elements. Denote the *type* of agent $i \in \{1, \ldots, n\}$ by t_i. Let T_i be the *type space* of agent i; that is, the possible types that agent i might have. Type spaces T_i can be arbitrary sets. Agent i's preferences over outcomes are modeled by the *valuation function* $v_i : A \times T_i \to \mathbb{R}$, where $v_i(a, t_i)$ is the valuation of agent i for outcome a when she has type t_i. Let us give three illustrative examples of this rather abstract and general definition.

Example 1 Consider the classical situation where the agents are n bidders in a *private value single-item auction*. Here, the mechanism designer is the auctioneer, and each agent i's type is described by the maximum amount she would be willing to pay for the indivisible item, say $t_i \in \mathbb{R}_{\geq 0}$. The $n+1$ possible outcomes A are then all possible assignments of the item to the agents, namely, $a \in \{0, 1\}^n$ so that $\sum_{i=1}^{n} a_i \leq 1$, meaning that agent i gets the item iff $a_i = 1$. The valuation functions are $v_i(a, t_i) = t_i$ if $a_i = 1$ and $v_i(a, t_i) = 0$ otherwise. If randomized allocations are allowed, we get $a \in [0, 1]^n$ instead of $a \in \{0, 1\}^n$, and as a general expression for the valuations, $v_i(a, t_i) = a_i t_i$. Note that the underlying combinatorial optimization problem to maximize the total valuation is trivial: choose the maximum among all t_i's and give the item to that bidder. The problem owes its interest only to the fact that the types t_i are private information of the bidders.

Example 2 An extension of Example 1 is an auction with a finite set of m heterogenous and indivisible items. Agent types $t_i \in \mathbb{R}_{\geq 0}^m$ denote values for each item, with the value for bundle $B \subseteq \{1, \ldots, m\}$ being equal to $\sum_{j \in B} t_{ij}$. The outcome space is then all partitions of m items into $n+1$ bundles, including bundles that stay at the auctioneer. Allowing even for arbitrary values for bundles rather than additive valuations, the problem is called a *combinatorial auction*. It has the same outcome space, but a richer set of types. For combinatorial auctions, the underlying optimization problem is the set packing problem, and generally NP-hard.

Example 3 Consider the following *demand rationing problem*. The mechanism designer is asked to distribute one unit of a divisible good amongst the agents, and the type of each of the n agents is her minimal demand $t_i \in (0, 1]$ for the good. In this setting, the outcomes A are all possible fractional distributions of the good over the agents. That is, all vectors $a \in [0, 1]^n$ so that $\sum_{i=1}^n a_i \leq 1$. The valuation functions are $v_i(a, t_i) = \min\{0, a_i - t_i\}$. Note that the valuations are either zero or negative and express the amount of demand rationing that an agent suffers. Here again, the underlying optimization problem to maximize the total valuation is almost trivial: if the demand exceeds one unit, any allocation of one unit is optimal, as long as no agent i receives more than her demand t_i. Otherwise, assign to every agent at least her demand. Again, the problem gets interesting as soon as the demands are private information.

Observe that the outcome spaces of the discrete version of the first and second example are finite, while it is infinite in the third. Also observe that in the first and the third example, the types of an agent i are single dimensional in the sense that $t_i \in \mathbb{R}$, while in the second example it is multi-dimensional. The above examples address utility maximization, but of course other objectives are possible as well. All that follows is general enough to cover these cases, too.

1.2 Mechanisms and Incentive Compatibility

The mechanism design problem we want to address here involves money to set incentives. More specifically, the task will be to find a mechanism that defines which allocation to choose and how much the agents need to pay for it. Payments could be both positive or negative. Formally, a *mechanism* (f, π) consists of an *allocation rule* $f : \prod_{i=1}^n T_i \to A$ and a *payment rule* $\pi : \prod_{i=1}^n T_i \to \mathbb{R}^n$. In fact what we describe here is a *direct revelation mechanism*, meaning that the only action of any agent is to reveal her type t_i. The allocation rule chooses for a vector t of type reports an outcome $f(t)$, and the payment rule assigns a payment $\pi_i(t)$ to be made by agent i.

On the agents' side we assume *quasi-linear utilities*, that is, the *utility* of agent i when the reported type vector is t, and the outcome is $a = f(t)$, equals valuation minus payment, that is,

$$v_i(f(t), t_i) - \pi_i(t).$$

Agents are assumed to be utility maximizers. Hence an agent, once asked by the mechanism designer about her type t_i, could decide to misreport her true type and pretend to have another type $s_i \in T_i$ instead. Indeed, if that false report s_i would result in an outcome and payment that yields higher utility than reporting true type t_i, agent i would not be truthful about her type. Once a mechanism (f, π) is fixed, the agents effectively play a noncooperative game of incomplete information, in which each agent i's strategy is to choose a reported type s_i, given true type t_i, so as to maximize her utility. Incomplete information refers to the fact that neither the mechanism designer nor any agent knows the types of the (other) agents. This implies for agents that their actions in equilibrium should ideally be robust against this uncertainty about other agents' types.

For what follows, let us denote by (t_i, t_{-i}) the vector of type reports when i reports t_i and the other agents' reports are t_{-i}. The following definition expresses the requirement that truthfulness yields the maximal utility for any agent, independent of the reports of other agents. This exactly corresponds to a dominant strategy equilibrium in the just mentioned noncooperative game.

Definition 4 (*Dominant strategy incentive compatible*) A mechanism (f, π) is called *dominant strategy incentive compatible*, or *truthful* for short, if for every agent i, every type $t_i \in T_i$, all type vectors t_{-i} that the other agents could report and every type $s_i \in T_i$ that i could report instead of t_i:

$$v_i(f(t_i, t_{-i}), t_i) - \pi_i(t_i, t_{-i}) \geq v_i(f(s_i, t_{-i}), t_i) - \pi_i(s_i, t_{-i}).$$

If for allocation rule f there exists a payment rule π such that (f, π) is a dominant strategy incentive compatible mechanism, then f is called *implementable in dominant strategies*, in short *implementable*.

This is a strong requirement. We can motivate it as follows: when the mechanism designer's goal is to optimize some objective function that depends on the true types of agents, he should better know these types. From the agents' perspectives the equilibrium is desirable as it provides a high degree of robustness: independent of the other agents' types, truthfulness is a best action. Even more, being truthful is optimal independent of the other agents' types *and* reports. As a matter of fact, this restriction to truthful mechanisms is much less restrictive than it seems: The celebrated *revelation principle* states that *any* mechanism with corresponding strategies of the agents that form an ex-post equilibrium can be equivalently replaced by a direct revelation mechanism in which truthfulness is a dominant strategy equilibrium. Here an ex-post equilibrium refers to a strategy profile in an incomplete information game, in which no agent would have liked to deviate after he has learned the true type of the other agents, and thus is ex-post satisfied with her action. In that sense truthfulness is a requirement that can be made *without loss of generality*.

1.3 Agenda

A first question that comes up is this: Which of all possible allocation rules are actually implementable? In other words, is it possible to augment all possible allocation rules f by appropriate payments π so the result is a truthful mechanism?

A second question is this: Assume that we have an implementable allocation rule f, how much flexibility does the mechanism designer have with respect to the payments π? For example, which are the minimal or maximal payments that implement that allocation rule?

The second question hints to an important property of mechanisms that is known as *revenue* or *payoff equivalence*. It describes situations where the payment rule π of a mechanism is already uniquely defined by the choice of an allocation rule f, up to adding constants. Let us formally define what it is.

Definition 5 (*Revenue equivalence*) An allocation rule f that is implementable satisfies the *revenue equivalence* property if for any two dominant strategy incentive compatible mechanisms (f, π) and (f, π') and any agent i there exist constants $h_i(t_{-i})$ that only depend on the reported types of the other agents, t_{-i}, such that

$$\forall\, t_i \in T_i : \pi_i(t_i, t_{-i}) = \pi'_i(t_i, t_{-i}) + h_i(t_{-i}).$$

The main purpose of this chapter is to show how these two fundamental questions can be rather elegantly addressed by reformulating them in terms of shortest path problems in graphs. Moreover, we hope that the reader who finds the previous concepts hard to digest, will find the graph-theoretic approach helpful and may even start to appreciate some of these concepts.

2 The Type Graph

Let us fix agent i and the reports, t_{-i}, of the other agents. For simplicity of notation we write $f(t_i)$ instead of $f(t_i, t_{-i})$, and $\pi(t_i)$ instead of $\pi_i(t_i, t_{-i})$, whenever t_{-i} is clear from the context.

For any allocation rule f, agent i and types of the other agents, t_{-i}, we define a *type graph* $G_f = G_f(i, t_{-i})$ as follows.[1] It has node set T_i and contains an arc from any node $s_i \in T_i$ to any other node $t_i \in T_i$ of length

$$\ell(s_i, t_i) = v_i(f(t_i), t_i) - v_i(f(s_i), t_i).$$

Note that there is an arc between any two nodes, as depicted in Fig. 1. The type graph is a complete, directed, and possibly infinite graph.

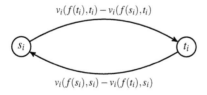

Fig. 1 Arc lengths of the type graph G_f between any two types $s_i, t_i \in T_i$ for agent i

---
[1] The type graph depends on the agent i and reports t_{-i} of the other agents. In order to keep notation simple, we will suppress the dependence on i and t_{-i}, and simply write G_f.

Here, $\ell(s_i, t_i)$ represents the gain in valuation for an agent truthfully reporting type t_i instead of lying some other type s_i. This gain could be positive or negative. A *path* from node s_i to node t_i in G_f, or (s_i, t_i)-path for short, is defined as $P = (s_i = t_{i_0}, t_{i_1}, \ldots, t_{i_k} = t_i)$ such that $t_{i_j} \in T_i$ for $j = 0, \ldots, k$. Denote by $length(P)$ the length of this path. A *cycle* is a path with $s_i = t_i$. We assume that each node represents a path from itself to itself and define the length of this (t_i, t_i)-path to be 0. Define $\mathscr{P}(s_i, t_i)$ to be the set of all (s_i, t_i)-paths in G_f.

In order to rephrase the question of implementability of an allocation rule in graph-theoretic terms, we need to define distances in the type graph. We let

$$dist_{G_f}(s_i, t_i) := \inf_{P \in \mathscr{P}(s_i, t_i)} length(P).$$

Observe that we here take the infimum over a possibly infinite set of (s_i, t_i)-paths in case the type graph has infinitely many nodes. If type spaces are finite, this is just the ordinary definition of shortest paths. In general, $dist_{G_f}(s_i, t_i)$ could be unbounded. However, if G_f does not contain a cycle of negative length then $dist_{G_f}(s_i, t_i)$ is finite, because the length of any (s_i, t_i)-path is bounded from below by $-\ell(t_i, s_i)$. This *nonnegative cycle* property of type graphs G_f turns out to be necessary and sufficient for the implementability of an allocation rule f.

3 Implementability and Payment Rules

A characterization of implementable allocation rules follows from the fact that incentive compatible payment rules exactly correspond to *node potentials* in the type graphs G_f. Recall that a node potential in a directed graph is a node labeling that satisfies the triangle inequality for each arc. Node potentials can be defined if and only if a graph has no cycle of negative length. In the mechanism design literature, the resulting characterization is known as *cyclical monotonicity*.

Theorem 6 (Cyclical monotonicity) *An allocation rule f is implementable if and only if none of the type graphs G_f contains a cycle of negative length.*

Proof Fix agent i and any type vector of the other agents t_{-i}. Assume f is implementable, then incentive compatibility translates into $\pi(t_i) - \pi(s_i) \leq \ell(s_i, t_i)$, for any $s_i, t_i \in T_i$. Adding up these inequalities along the arcs of any cycle shows that the cycle has nonnegative length.

On the other hand, assume that G_f has no cycle of negative length. Then fix some $t_{i_0} \in T_i$, and observe that the distance function $dist_{G_f}(t_{i_0}, t_i)$ is well defined for all $t_i \in T_i$. Moreover, $\pi(t_i) := dist_{G_f}(t_{i_0}, t_i)$ defines an incentive compatible payment rule for f, since $dist_{G_f}(t_{i_0}, t_i) \leq dist_{G_f}(t_{i_0}, s_i) + \ell(s_i, t_i)$ for all $s_i, t_i \in T_i$. □

An immediate consequence of Theorem 6 is that an implementable allocation rule must satisfy the nonnegative cycle property for all cycles of length two, that is,

$\ell(s_i, t_i) + \ell(t_i, s_i) \geq 0$. This is known in the mechanism design literature as *(weak) monotonicity* of an implementable allocation rule. Due to its importance, we explicitly state this consequence as a separate theorem.

Theorem 7 (Weak monotonicity) *If an allocation rule f is implementable, then for every i and t_{-i}, the following (weak) monotonicity condition must be satisfied: For all $s_i, t_i \in T_i$*

$$(v_i(f(t_i), t_i) - v_i(f(t_i), s_i)) + (v_i(f(s_i), s_i) - v_i(f(s_i), t_i)) \geq 0. \tag{1}$$

For example, in the single-item auction setting from Example 1 where $t_i \in \mathbb{R}_{\geq 0}$ and $v_i(f(t_i), t_i) = t_i \cdot f_i(t_i)$, inequality (1) is equivalent to $(f_i(t_i) - f_i(s_i))(t_i - s_i) \geq 0$. In other words, for any implementable mechanism the probability of allocating the item to a bidder, expressed by $f_i(t_i)$, must be monotonically non-decreasing in the valuation that the bidder has for the item.

The proof of Theorem 6 suggests even more: incentive compatible payment rules for an implementable allocation rule f can be defined by computing shortest paths in type graphs. Let us push this idea a little further. For each agent i we choose the payment at some type $t_{i_0} \in T_i$ arbitrarily,[2] say $\pi_i(t_{i_0}) = 0$. Then, in i's type graph G_f, we compute shortest path lengths $dist_{G_f}(t_{i_0}, t_i)$ to all types $t_i \in T_i$. This is well defined by Theorem 6. From the proof of Theorem 6, it follows that an incentive compatible payment rule can be defined by letting

$$\pi^+(t_i) := dist_{G_f}(t_{i_0}, t_i)$$

for all $t_i \in T_i$. But we can also compute shortest path lengths from all types $t_i \in T_i$ to t_{i_0}, which yields another incentive compatible payment rule, namely

$$\pi^-(t_i) := -dist_{G_f}(t_i, t_{i_0})$$

for all $t_i \in T_i$. Indeed, incentive compatibility for the latter follows because for all $s_i, t_i \in T_i, dist_{G_f}(s_i, t_{i_0}) \leq \ell(s_i, t_i) + dist_{G_f}(t_i, t_{i_0})$, and thus $-dist_{G_f}(t_i, t_{i_0}) \leq -dist_{G_f}(s_i, t_{i_0}) + \ell(s_i, t_i)$. The following lemma relates these payment rules to *any* incentive compatible payment rule π.

Lemma 8 *Let π be any payment rule that implements allocation rule f, and assume w.l.o.g.[3] that $\pi(t_{i_0}) = 0$ for $t_{i_0} \in T_i$ and all i, then*

$$\pi^- \leq \pi \leq \pi^+.$$

---

[2] For example in auction settings $\pi(t_{i_0}) = 0$ is a natural choice for any type t_{i_0} for which bidder i does not win anything in the auction.

[3] This is indeed no loss of generality as we can replace $\pi_i(\cdot)$ by $\pi_i(\cdot) - \pi_i(t_{i_0})$ otherwise, for all i.

Proof Consider agent i. Observe that, for any $s_i, t_i \in T_i$, by adding up the incentive compatibility constraints of π along any (s_i, t_i)-path P in G_f, we have $\pi(t_i) \leq \pi(s_i) + length(P)$. Taking the infimum over all such paths in $\mathscr{P}(s_i, t_i)$, we get

$$\pi(t_i) \leq \pi(s_i) + dist_{G_f}(s_i, t_i). \tag{2}$$

Therefore, letting $s_i = t_{i_0}$ in (2) we get $\pi(t_i) = \pi(t_i) - \pi(t_{i_0}) \leq dist_{G_f}(t_{i_0}, t_i) = \pi^+(t_i)$. On the other hand, letting $t_i = t_{i_0}$ and $s_i = t_i$ in (2), we see that $\pi(t_i) = \pi(t_i) - \pi(t_{i_0}) \geq -dist_{G_f}(t_i, t_{i_0}) = \pi^-(t_i)$. □

Equipped with Lemma 8 we get a characterization of implementable allocation rules that satisfy revenue equivalence.

Theorem 9 (Characterization of revenue equivalence) *Let f be an allocation rule that is implementable. Then f satisfies revenue equivalence if and only if in all type graphs G_f the distances are anti-symmetric, that is, $dist_{G_f}(s_i, t_i) = -dist_{G_f}(t_i, s_i)$ for all $s_i, t_i \in T_i$.*

Proof Let f satisfy revenue equivalence. Fix an agent i and a report vector t_{-i} of the other agents, and let G_f be i's type graph. Let $s_i, t_i \in T_i$. Then the functions $dist_{G_f}(s_i, \cdot)$ and $dist_{G_f}(t_i, \cdot)$ define both incentive compatible payment rules for f. By revenue equivalence they differ only by a constant $h(t_{-i})$. Hence we have that $dist_{G_f}(s_i, \cdot) - dist_{G_f}(t_i, \cdot) = h(t_{-i})$. Especially, if we plug s_i and t_i into this equation we get that $dist_{G_f}(s_i, s_i) - dist_{G_f}(t_i, s_i) = dist_{G_f}(s_i, t_i) - dist_{G_f}(t_i, t_i)$. As $dist_{G_f}(s_i, s_i) = dist_{G_f}(t_i, t_i) = 0$, we see $dist_{G_f}(s_i, t_i) = -dist_{G_f}(t_i, s_i)$.

For the reverse direction, take an implementable allocation rule f and suppose that $dist_{G_f}(s_i, t_i) = -dist_{G_f}(t_i, s_i)$ for all $s_i, t_i \in T_i$. Take any incentive compatible payment rule π and a type $t_{i_0} \in T_i$. By $dist_{G_f}(t_{i_0}, \cdot) = -dist_{G_f}(\cdot, t_{i_0})$ and Lemma 8 we get $\pi(\cdot) - \pi(t_{i_0}) = \pi^-(\cdot) = \pi^+(\cdot) = dist_{G_f}(t_{i_0}, \cdot)$. Thus for any two payment rules π and π' we get $\pi(\cdot) - \pi(t_{i_0}) = \pi'(\cdot) - \pi'(t_{i_0})$, which proves revenue equivalence by letting $h(t_{-i}) := \pi(t_{i_0}) - \pi'(t_{i_0})$. □

We have seen until here that, up to normalization at some type t_{i_0}, there is a minimal and a maximal payment rule, and revenue equivalence holds when these two coincide. But there is even more we can say about the space of payment rules, and it is surprisingly simple. We need two more definitions. For two payment rules π and π' that implement an allocation rule f, define

$$\hat{\pi}(t_i) := \max\{\pi(t_i), \pi'(t_i)\} \quad \text{and} \quad \check{\pi}(t_i) := \min\{\pi(t_i), \pi'(t_i)\}$$

as the *join* and *meet* of π and π'. Then these payment rules are incentive compatible, too. To conveniently formulate this, let us again assume w.l.o.g. that all payment rules are normalized at some t_{i_0} for all agents i, so that $\pi(t_{i_0}) = 0$.

Theorem 10 (Payment rules form a lattice) *Consider an implementable allocation rule f. Then the set of all payment rules that implement f define a bounded lattice* with respect to the meet and join definition. The payment rules π^- and π^+ are the minimal, respectively maximal elements of that lattice.

Proof We are only left to prove that join $\hat{\pi}$ and meet $\check{\pi}$ are both incentive compatible payments for f. This follows immediately from the fact that payment rules correspond to node potentials in the type graphs, and it is well known that the set of node potentials in a directed graph forms a lattice. Let us give the simple proof here: Take any two $s_i, t_i \in T_i$, and first consider $\hat{\pi}$. Say w.l.o.g. that $\hat{\pi}(t_i) = \pi(t_i)$. Then, as π is incentive compatible, $\hat{\pi}(t_i) = \pi(t_i) \leq \pi(s_i) + \ell(s_i, t_i) \leq \hat{\pi}(s_i) + \ell(s_i, t_i)$. Likewise, for $\check{\pi}$, assume w.l.o.g. that $\check{\pi}(s_i) = \pi(s_i)$, then, as π is incentive compatible, $\check{\pi}(s_i) + \ell(s_i, t_i) = \pi(s_i) + \ell(s_i, t_i) \geq \pi(t_i) \geq \check{\pi}(t_i)$. □

4 Theorems at Work

We here sketch how the theorems just derived can help to derive qualitative insights into some relevant problems.

4.1 Implementations for Demand Rationing

Recall the demand rationing problem introduced in Example 3. One possible allocation rule f would be the *dictatorial rule*: Choose some agent, say agent 1 and give her whatever she demands, $a_1 = f_1(t) := t_1$. Divide the remainder equally over the remaining agents, $a_i = f_i(t) := (1 - t_1)/(n - 1)$. This allocation rule does not maximize total utility when $f_i(t) > t_i$ for some i, but this is not essential for what follows. We can use Theorems 6 and 9 to address implementability and revenue equivalence. First, since the outcome is independent of the reports of agents $2, \ldots, n$, their type graphs have arcs of length zero, and truthfulness is trivially a dominant strategy for them as long as their payment is constant over all types. Now focus on agent 1, and let us drop index 1 in the following. Denote by $s < t$ two possible types in T_1, then the arc lengths in the type graph G_f for agent 1 are

$$\ell(s, t) = v(f(t), t) - v(f(s), t) = v(t, t) - v(s, t) = 0 - (s - t) = t - s > 0, \text{ and}$$
$$\ell(t, s) = v(f(s), s) - v(f(t), s) = v(s, s) - v(t, s) = 0 - 0 = 0.$$

As all arc lengths in all type graphs have nonnegative length, f is implementable by Theorem 6. Moreover, for any three types $s < q < t$ we have that $\ell(s, q) + \ell(q, t) = (q - s) + (t - q) = t - s = \ell(s, t)$; see Fig. 2.

It follows that the length of *any* (s, t)-path is at least $t - s$, and hence $dist_{G_f}(s, t) = \ell(s, t) = t - s$. Therefore, $dist_{G_f}(s, t) + dist_{G_f}(t, s) > 0$, and by Theorem 9 allocation rule f does not qualify for revenue equivalence. Indeed, if we normalize the payments at $\pi_i(1) = 0$ for all agents i, then the maximal incentive

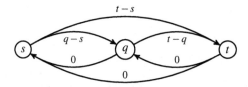

Fig. 2 Demand rationing: part of the type graph G_f for agent 1 for the dictatorial allocation rule

compatible payment for agent 1 is $\pi_1^+(t_1) = 0$ for all $t_1 \in (0, 1]$, and the minimal incentive compatible payment is $\pi_1^-(t_1) = t_1 - 1$ for all $t_1 \in (0, 1]$. These payments are exactly the negative of the shortest path lengths from type $t_{1_0} = 1$ to all other types t_1 in the former case, and from types t_1 to type $t_{1_0} = 1$ in the latter case. (Payments for all other agents being constant and equal to zero.)

Another, more reasonable allocation rule for this demand rationing game is the *proportional allocation rule*: define outcome $a_i = f_i(t) := t_i / \sum_{k=1}^n t_k$ for all agents i. This allocation rule is implementable as well, and it turns out to have a unique payment rule by revenue equivalence. Just like for the dictatorial allocation rule, this can be shown by verifying cyclical monotonicity as well as the antisymmetry of distances in the corresponding type graphs, using Theorems 6 and 9. Even though this is basic, it turns out to be a bit tedious and we do not work it out here. The reason is that the description of arc lengths in the type graph involves several case distinctions, such as the case where the total demand of all agents except i exceeds 1, i.e. $\sum_{k \neq i} t_k \geq 1$, and the case where it does not.

The demand rationing example owes its particular interest to the fact that many known theorems about revenue equivalence remain silent on it. The reason is that these results usually give characterizations of type spaces T and/or valuations v that allow to conclude revenue equivalence for *all* implementable allocation rules. But here we have two implementable allocation rules, one of which fulfills revenue equivalence, while the other does not.

4.2 Uniqueness of the Vickrey Auction for Single-Item Auctions

Consider the single-item auction setting from Example 1, and recall the Vickrey auction: it collects a bid from each bidder, allocates the item to a bidder with highest bid, and charges this bidder a price equal to the highest bid of a loosing bidder. We here show that the Vickrey auction is in fact the unique incentive compatible single-item auction that allocates the item to the bidder with highest valuation.

Fix agent i and t_{-i}, and again let us drop index i for simplicity. Consider *any* implementable deterministic allocation rule f. By Theorem 7 the allocation rule must be monotone. This implies that f allocates the item to agent i either never, or

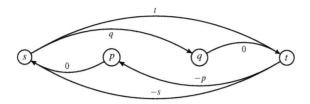

Fig. 3 Single-item auction: part of the type graph G_f for agent i, q being i's threshold value

always, or there exists a threshold type q, such that $f(t) = 0$ for $t < q$ and $f(t) = 1$ for $t > q$. Let us assume $f(q) = 1$. (The case $f(q) = 0$ works analogously.)

Let $s < t$ be two types of agent i. Note that $\ell(s, t) = 0$ if $t < q$ or $s \geq q$. Otherwise,

$$\ell(s, t) = t \quad \text{and} \quad \ell(t, s) = -s.$$

This implies that, for $s < q < t$, arc (s, t) with length t can be made shorter by replacing it by the two arcs $(s, q), (q, t)$ with total length q. And any arc (t, s) can be made shorter by replacing it by the two arcs $(t, p), (p, s)$, with $s < p < q$, with total length $-p$; see Fig. 3. Using this observation, we can argue that all cycles have nonnegative length. Therefore, by Theorem 6 every deterministic, monotone allocation rule f is implementable. By taking the limit $p \to q$, it also follows that

$$dist_{G_f}(s, t) = \begin{cases} 0 & \text{if } t < q, \\ q & \text{if } s < q \leq t, \\ 0 & \text{if } s \geq q, \end{cases} \quad \text{and} \quad dist_{G_f}(t, s) = \begin{cases} 0 & \text{if } t < q, \\ -q & \text{if } s < q \leq t, \\ 0 & \text{if } s \geq q. \end{cases}$$

In particular, we get $dist_{G_f}(s, t) = -dist_{G_f}(t, s)$. Hence, by Theorem 9, there is a unique payment rule for implementing f, up to constants.

It is a reasonable requirement that a bidder who does not win the item pays zero. This fixes the payment at bids strictly lower than q to zero. Now consider the allocation rule that assigns to the bidder with highest valuation, say i. It must have a threshold value q_i equal to the highest $t_j, j \neq i$, in other words the second highest bid. Revenue equivalence tells us that charging the winner the second highest bid is the only way of making this allocation rule dominant strategy incentive compatible.

Theorem 11 (Uniqueness of Vickrey auction) *The Vickrey auction is the unique direct, deterministic, dominant strategy incentive compatible auction that assigns the item to the bidder with highest valuation and charges losers nothing.*

5 Discussion and Literature

Characterizing implementable allocation rules by cyclical monotonicity as in Theorem 6 goes back to a paper by Rochet [10]; the term cyclical monotonicity is from convex analysis [11]. The graph-theoretic view and characterization of implementable allocation rules was first developed by Gui et al. [5], and has been fully explored by Vohra in [13]. There is a series of papers that study situations where the necessary weak monotonicity condition of Theorem 7 suffices to guarantee implementability, going far beyond the single-item auction setting illustrated in Sect. 4. Saks and Yu [12] proved it for convex type spaces, extending partial results by Bikhchandani et al. [3] and [5]. Berger et al. [2] further generalized these results. For implementation in Bayes–Nash equilibrium with single-dimensional type spaces this result goes back to Myerson [9]. Ashlagi et al. [1] have shown that convexity of the type space is essentially necessary. The demand rationing problem in Example 3 is from [4], and the valuation functions also appear in [7]. For the revelation principle in mechanism design, and also a version of the revenue equivalence theorem, see the seminal paper by Myerson on the design of optimal auctions [9]. The graph-theoretic characterization of revenue equivalence presented in Theorem 9 is taken from Heydenreich et al. [6]. Kos and Messner [8] extended this result by identifying the lattice structure of incentive compatible payments. Theorem 11 can be extended to many multi-dimensional domains, showing that any implementable allocation rule satisfies revenue equivalence and thereby proving that the multi-dimensional generalization of the Vickrey auction, called the Vickrey–Clarke–Groves (VCG) mechanism, is the only way to implement the utilitarian maximizer. The graph-theoretic approach yields elegant proofs of such results.

References

1. Ashlagi, I., Braverman, M., Hassidim, A., Monderer, D.: Monotonicity and implementability. Econometrica **78**(5), 1749–1772 (2010)
2. Berger, A., Müller, R., Naeemi, S.H.: Characterizing implementable allocation rules in multi-dimensional environments. Memorandum RM/14/021, Maastricht University (2014)
3. Bikhchandani, S., Chatterji, S., Lavi, R., Mu'alem, A., Nisan, N., Sen, A.: Weak monotonicity characterizes deterministic dominant-strategy implementation. Econometrica **74**(4), 1109–1132 (2006)
4. Cachon, G., Lariviere, M.: Capacity choice and allocation: strategic behavior and supply chain performance. Manag. Sci. **45**(8), 1091–1108 (1999)
5. Gui, H., Müller, R., Vohra, R.: Dominant strategy mechanisms with multidimensional types. Discussion Paper 1392, Northwestern University (2004)
6. Heydenreich, B., Müller, R., Uetz, M., Vohra, R.: Characterization of revenue equivalence. Econometrica **77**(1), 307–316 (2009)
7. Holmström, B.: Groves' scheme on restricted domains. Econometrica **47**(5), 1137–1144 (1979)
8. Kos, N., Messner, M.: Extremal incentive compatible transfers. J. Econ. Theory **148**(1), 134–164 (2012)
9. Myerson, R.B.: Optimal auction design. Math. Oper. Res. **6**(1), 58–73 (1981)
10. Rochet, J.-C.: A necessary and sufficient condition for rationalizability in a quasi-linear context. J. Math. Econ. **16**(2), 191–200 (1987)
11. Rockafellar, R.T.: Convex Analysis. Princeton University Press, Princeton (1970)
12. Saks, M., Yu, L.: Weak monotonicity suffices for truthfulness on convex domains. In: Proceedings of the 6th ACM Conference on Electronic Commerce, pp. 286–293. ACM (2005)
13. Vohra, R.: Mechanism Design: A Linear Programming Approach. Econometric Society Monographs. Cambridge University Press, Cambridge (2011)

Selfish Routing and Proportional Resource Allocation

A Joint Bound on the Loss of Optimality

Andreas S. Schulz

Abstract We consider two optimization problems, multicommodity flow and resource allocation, from a game-theoretic point of view. We show that two known bounds on the respective losses of optimality can be derived from the same geometric quantity, which yields a simple proof of either result.

1 Introduction

In several applications of multicommodity flow, it may be impractical or outright impossible to centrally control the entire network. The observed flow may rather be the consequence of the actions of many individual agents, which may each have their own optimality criteria. It is a central result in algorithmic game theory that the loss of optimality in multicommodity flow networks due to selfish behavior of the participating agents is bounded from above by 4/3, if the cost of each arc depends linearly on the flow of that arc.

In a different context, competing agents with quasilinear payoffs may bid for their share of a divisible resource (e.g., bandwidth). One way of allocating that resource is to give each bidder a fraction proportional to their bid and ask them to pay the bid. While the outcome of this rule may not maximize the total utility, the deviation from optimality can also be bounded from above by 4/3, if the bidders' utility functions are concave.

The latter result was obtained shortly after and independently of the 4/3 bound for selfish routing in multicommodity flow networks. The original proofs were quite different from one another, and the question was quickly raised whether there might be a connection between these two results. (Note that either bound is tight.)

In this chapter, we present short proofs for both results. In fact, we will show that the core argument of either proof relies on the same geometric insight. This enables us to give a unified presentation of both proofs, which may serve as evidence of some relationship between both models.

A.S. Schulz (✉)
Fakultät für Mathematik und Fakultät für Wirtschaftswissenschaften, Technische Universität München, Arcisstr. 21, 80333 München, Germany
e-mail: andreas.s.schulz@tum.de

2 Selfish Routing

An instance of the multicommodity flow problem considered here is given by a network with arc set A and flow-dependent arc costs $c_a(\cdot)$. We assume that $c_a : \mathbb{R}_+ \to \mathbb{R}_+$ is a non-decreasing and continuous function, for all $a \in A$. Additionally, there are a number of source-sink pairs, where the source (sink) of pair k has supply (demand) d_k. Viewed from an optimization perspective, the objective is to find a flow that satisfies supply and demand at minimal cost. In mathematical terms,

$$\min \sum_{a \in A} c_a(x_a) \, x_a$$

$$\text{s.t.} \sum_{P \ni a} x_P = x_a \quad \text{for all arcs } a \in A,$$

$$\sum_{P \in \mathcal{P}_k} x_P = d_k \quad \text{for all source-sink pairs } k,$$

$$x_P \geq 0 \quad \text{for all paths } P \in \mathcal{P}.$$

Here, x_a denotes the (amount of) flow on arc a, $\mathcal{P}_k$ is the set of all paths between the source and sink of pair k, $\mathcal{P} = \cup_k \mathcal{P}_k$, and x_P is the flow on path P.

From a game-theoretic point of view, a source-sink pair k corresponds to a continuum $(0, d_k]$ of selfish agents, each of which controls an infinitesimally small amount of flow. Agents choose their routes independently, so as to minimize their individual cost, where the cost of path P under flow x is equal to $c_P(x) = \sum_{a \in P} c_a(x_a)$. Hence, in equilibrium, no agent can improve their cost by changing their path. Put differently, x^{EQ} is an equilibrium flow if for all source-sink pairs k and any pair of paths $P, Q \in \mathcal{P}_k$ with $x_P^{\text{EQ}} > 0$, we have $c_P(x^{\text{EQ}}) \leq c_Q(x^{\text{EQ}})$. It follows that an equilibrium flow, x^{EQ}, satisfies

$$\sum_{P \in \mathcal{P}} c_P(x^{\text{EQ}}) \, x_P^{\text{EQ}} \leq \sum_{P \in \mathcal{P}} c_P(x^{\text{EQ}}) \, x_P \tag{1}$$

for all feasible (path) flows x. Equivalently, by the standard transformation between path flow and arc flow,

$$\sum_{a \in A} c_a(x_a^{\text{EQ}}) \, x_a^{\text{EQ}} \leq \sum_{a \in A} c_a(x_a^{\text{EQ}}) \, x_a \tag{2}$$

for all feasible (arc) flows x.[1]

---

[1] Note that in (1) and (2), the cost of each path and arc, respectively, is fixed in advance according to its load experienced in some equilibrium. Such equilibrium costs are, in fact, unique, even if instances admit more than one equilibrium flow.

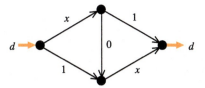

Fig. 1 An instance of the multicommodity flow problem with unique source-sink pair. Arcs are labeled with their respective cost functions. Demand is equal to $d = 1$. It is optimal to send half of the flow on the upper path and the other half on the lower path, for a total cost of $3/2$. The (unique) equilibrium is to send all flow along the path consisting of three arcs. The corresponding cost is 2.

The example in Fig. 1 shows that the total cost in equilibrium may exceed that of an optimal flow, even if the cost of each arc depends linearly on the flow of that arc. Nonetheless, with the help of (2), the loss of optimality can be bounded, as stated in the first main result of this chapter.

Theorem 1 *The cost of any equilibrium in a multicommodity flow network with linear cost functions is at most $4/3$ that of an optimal flow.*

We will prove Theorem 1 in Sect. 4, together with a similar result for a very different situation, which is described in the next section.

3 Proportional Resource Allocation

An instance of the proportional resource allocation model is given by a single, infinitely divisible resource of unit capacity and a set of $n \geq 2$ players, competing for that resource. Each player has their own, strictly increasing, concave and continuously differentiable, utility function $U_i : [0, 1] \to \mathbb{R}_+$. The proportional sharing protocol allocates the resource as follows. Player i bids some nonnegative amount b_i and, in return, obtains the fraction $x_i = b_i / \sum_{j=1}^{n} b_j$ of the resource. The payment charged to Player i is equal to their bid, and so their payoff equals $U_i(x_i) - b_i$. In equilibrium, each player bids so as to maximize their payoff, given the bids of the other players. Using single-variable calculus, and $B = \sum_{i=1}^{n} b_i$, it is straightforward to see that a bid vector $b = (b_i)$ is in equilibrium if and only if, for all players i,

$$(1 - x_i)U_i'(x_i) = B \text{ if } b_i > 0 \text{ and } U_i'(0) \leq B \text{ if } b_i = 0. \tag{3}$$

As in selfish routing, the resource allocation corresponding to an equilibrium bid may not maximize the total utility, $\sum_{i=1}^{n} U_i(x_i)$, as shown by Example 2.

Example 2 Player 1 has utility function $U_1(x) = x$. For every other player, $i = 2,\ldots, n$, $U_i(x) = x/2$. It would be optimal to allocate the entire resource to Player 1.[2] A little bit of calculus shows that, in equilibrium, Player 1 bids $b_1 = nB/(2n-1)$, whereas any other player bids $b_i = B/(2n-1)$. As the number of players grows large, Player 1 obtains about half of the resource; the rest is split equally among the other players. The total utility of that allocation is approximately $3/4$.

However, as we will see, (3) implies that the resource allocation vector x^{EQ} corresponding to an equilibrium bid vector b^{EQ} is an optimal solution of the following concave maximization problem:

$$\max \quad \sum_{i=1}^{n}(1 - x_i^{EQ})U_i(x_i) \tag{4a}$$

$$\text{s.t.} \quad \sum_{i=1}^{n} x_i \leq 1, \tag{4b}$$

$$x_i \geq 0 \quad \text{for } i = 1,\ldots, n. \tag{4c}$$

Here, it is important to note that x^{EQ} is part of the input; the variables are the x_i.[3] Using dual variables λ for (4b) and μ_i for (4c), the Karush-Kuhn-Tucker optimality conditions for (4) are

$$\begin{aligned}
(1 - x_i^{EQ})U_i'(x_i) &= \lambda + \mu_i & \text{for } i = 1,\ldots, n, \\
\mu_i x_i &= 0 & \text{for } i = 1,\ldots, n, \\
\lambda(1 - \sum_{i=1}^{n} x_i) &= 0, & \\
\mu_i &\leq 0 & \text{for } i = 1,\ldots, n, \\
\lambda &\geq 0, &
\end{aligned}$$

and primal feasibility, of course. With $\lambda = B$ and $\mu_i = U_i'(0) - B$ if $x_i^{EQ} = 0$ (and $\mu_i = 0$ otherwise), it follows from (3) that x^{EQ} is indeed a maximum of (4). In particular, x^{EQ} satisfies

$$\sum_{i=1}^{n}(1 - x_i^{EQ})U_i(x_i) \leq \sum_{i=1}^{n}(1 - x_i^{EQ})U_i(x_i^{EQ}) \tag{5}$$

for all feasible allocations x. As will be demonstrated in Sect. 4, Eq. (5) can be used to prove the second main result of this chapter:

Theorem 3 *The total utility of an optimal resource allocation is at most $4/3$ that of an equilibrium allocation.*

---

[2]Note that this allocation cannot arise from an equilibrium; only the bid of Player 1 would be positive, who could bid less and still receive the entire resource.
[3]Similarly to the case of uncoordinated routing in Sect. 2, the equilibrium allocation is known to be unique.

4 The Proof

With (2) and (5) as starting points, we are able to give a simple, unified proof of Theorems 1 and 3. For an equilibrium flow $x^{EQ} = (x_a^{EQ})$ and an equilibrium allocation $x^{EQ} = (x_i^{EQ})$ as well as an arbitrary feasible flow $x = (x_a)$ and an arbitrary feasible allocation $x = (x_i)$, we need to show

$$\sum_{a \in A} c_a(x_a^{EQ}) x_a^{EQ} \leq \frac{4}{3} \sum_{a \in A} c_a(x_a) x_a \quad \text{and} \quad \sum_{i=1}^{n} U_i(x_i) \leq \frac{4}{3} \sum_{i=1}^{n} U_i(x_i^{EQ}),$$

respectively. We slightly rewrite both inequalities to allow for a more intuitive proof:

$$\sum_{a \in A} c_a(x_a^{EQ}) x_a^{EQ} - \sum_{a \in A} c_a(x_a) x_a \leq \frac{1}{4} \sum_{a \in A} c_a(x_a^{EQ}) x_a^{EQ}, \tag{6}$$

$$\sum_{i=1}^{n} U_i(x_i) - \sum_{i=1}^{n} U_i(x_i^{EQ}) \leq \frac{1}{4} \sum_{i=1}^{n} U_i(x_i). \tag{7}$$

Thus, our goal is to establish (6) and (7). By a small manipulation of (2) and (5), respectively, we obtain

$$\sum_{a \in A} c_a(x_a^{EQ}) x_a^{EQ} - \sum_{a \in A} c_a(x_a) x_a \leq \sum_{a \in A} \left(c_a(x_a^{EQ}) - c_a(x_a)\right) x_a \quad \text{and}$$

$$\sum_{i=1}^{n} U_i(x_i) - \sum_{i=1}^{n} U_i(x_i^{EQ}) \leq \sum_{i=1}^{n} \left(U_i(x_i) - U_i(x_i^{EQ})\right) x_i^{EQ}.$$

We can capitalize on the similarity of the expressions on the right-hand side by defining

$$y_1 := x_a, \quad y_2 := x_a^{EQ} \quad \text{and} \quad z(\cdot) := c_a(\cdot)$$

in the case of selfish routing, and

$$y_1 := x_i^{EQ}, \quad y_2 := x_i \quad \text{and} \quad z(\cdot) := U_i(\cdot)$$

for proportional resource allocation. Consequently, it suffices to prove

$$\left(z(y_2) - z(y_1)\right) y_1 \leq \frac{1}{4} z(y_2) y_2 \tag{8}$$

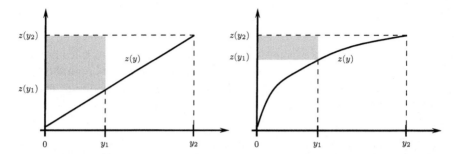

Fig. 2 The drawing on the *left* illustrates the situation for linear functions. The left-hand side of (8) corresponds to the area of the small, shaded rectangle, which is defined by the two opposite corners $(0, z(y_1))$ and $(y_1, z(y_2))$. This rectangle is completely contained inside the triangle defined by $(0, z(0))$, $(0, z(y_2))$, and $(y_2, z(y_2))$. In turn, this triangle is inscribed in the large rectangle defined by the origin and $(y_2, z(y_2))$ as opposite corners. The result follows from comparing the area of the small rectangle with that of the large rectangle. The figure on the *right* shows the same situation for an arbitrary concave function. Obviously, we obtain the linear case as the extreme case, which implies that the same bound holds for concave functions.

to obtain (6) and (7).[4] Since z is non-negative and non-decreasing and y_1 and y_2 are non-negative, it all comes down to verifying (8) for $y_1 < y_2$. It is enlightening to graph the situation, which is done in Fig. 2.

With the help of Fig. 2, it becomes clear that proving (8) reduces to showing that the area of the smaller, shaded rectangle is at most a quarter of that of the larger rectangle, which is defined by taking the origin and the point $(y_2, z(y_2))$ as opposite corners. That this is true follows immediately from two basic geometric facts:

- The area of any rectangle that fits inside a given triangle is at most half that of the triangle.
- The area of any triangle that fits inside a given rectangle is at most half that of the rectangle.

This not only completes the proof of Theorems 1 and 3, but also shows that Theorem 1 holds for concave functions c_a as well:

Theorem 4 *The cost of any equilibrium in a multicommodity flow network with concave cost functions is at most 4/3 that of an optimal flow.*

The main lesson of this chapter is that the same geometric quantity determines the worst-case bound of selfish routing and proportional resource allocation.

---

[4]For (7), we also use that $x_i \leq 1$ and that U_i is non-negative for all i, which implies that $\sum_{i=1}^{n} U_i(x_i)x_i \leq \sum_{i=1}^{n} U_i(x_i)$.

5 Chapter Notes

The equilibrium concept considered in the context of the multicommodity flow problem is known as Wardrop equilibrium [22] or user equilibrium [8]. It is a basic tool in transportation science, used to describe commuter behavior in real-life networks [9]. Beckmann et al. [1] proved that an equilibrium flow always exists, under the assumptions stated in Sect. 2. It also follows from their work that an equilibrium can be computed using convex optimization techniques. The variational inequality characterization, i.e., (1) or (2), was given in [21]; see also [7]. A Wardrop equilibrium can be seen as the limit of a Nash equilibrium [15] in a game with a finite number of players, when the number of players goes to infinity [11]. For more information and references on Wardrop equilibria, the reader is referred to [6].

The fact that equilibria (in general) and Wardrop equilibria (in particular) need not be optimal was noticed early on, a phenomenon known in economics as the loss of efficiency. The example in Fig. 1 is a slightly extended version of an instance that appeared in [17]; see also [2]. The idea to analyze equilibria from a worst-case perspective was put forward in [13]; the worst-case ratio between the value of an equilibrium and that of an optimum is now widely known as the "price of anarchy" [16]. Theorem 1, which helped to define the field of algorithmic game theory, was originally proved in [19]. The geometric proof presented here is derived from [4] and was inspired by an earlier proof by the same authors [3]. The latter paper also contains the result stated in Theorem 4.

The price of anarchy of the proportional allocation mechanism was analyzed in [12], where Theorem 3 was proved. The same source also contains references and a proof, which show that the equilibrium considered here, a Nash equilibrium, exists and the resulting allocation is unique (under the assumptions stated in Sect. 3). [18] contains another proof of Theorem 3. However, the proof presented here and the relation to the proof of Theorem 1 in [4] was first given in [5], including the characterization provided by (5). The question whether there is a relationship between the two models considered here was already raised in [12, p. 418]. The proof of Sect. 4 may suggest that there is a broader class of settings where this bound applies. It remains an interesting open question to characterize the underlying commonality that makes this approach work.

The bounds presented in Theorems 1 and 3 are tight, as demonstrated by the instances given in Fig. 1 and Example 2, respectively. Example 2 is lifted from [12]. It is discussed, in more detail, in [20, p. 453, Ex. 17.6].

References [14] and [10] contain accounts of the history and pointers to various proofs of the first geometric fact used above. The situation here is actually a bit simpler than the general case because the triangle in Fig. 2 is a right triangle and two sides of the small rectangle lie on the legs of the triangle. The same is true for the triangle and the large rectangle, which renders the second fact trivial.

References

1. Beckmann, M.J., McGuire, C.B. Winsten, C.B.: Studies in the Economics of Transportation. Yale University Press (1956)
2. Braess, D.: Über ein Paradoxon aus der Verkehrsplanung. Unternehmensforschung **12**, 258–268 (1969). English translation: Braess, D., Nagurney, A., Wakolbinger, T.: On a paradox of traffic planning. Trans. Sci. **39**, 446–450 (2005)
3. Correa, J.R., Schulz, A.S., Stier-Moses, N.E.: Selfish routing in capacitated networks. Math. Oper. Res. **29**, 961–976 (2004)
4. Correa, J.R., Schulz, A.S., Stier-Moses, N.E.: A geometric approach to the price of anarchy in nonatomic congestion games. Games Econ. Behav. **64**, 457–469 (2008)
5. Correa, J.R., Schulz, A.S., Stier-Moses, N.E.: The price of anarchy of the proportional allocation mechanism revisited. In: Chen, Y., Immorlica, N. (eds.) Web and Internet Economics. Lecture Notes in Computer Science, Vol. 8289, pp. 109–120. Springer (2013)
6. Correa, J.R., Stier-Moses, N.E.: Wardrop equilibria. In: Wiley Encyclopedia of Operations Research and Management Science. Wiley (2011)
7. Dafermos, S.C.: Traffic equilibrium and variational inequalities. Trans. Sci. **14**, 42–54 (1980)
8. Dafermos, S.C., Sparrow, F.T.: The traffic assignment problem for a general network. J. Res. U.S. Nat. Bur. Stand. **73B**, 91–118 (1969)
9. Florian, M.: Untangling traffic congestion: Application of network equilibrium models in transportation planning. OR/MS Today **26**, 52–57 (1999)
10. Gardner, M.: Some surprising theorems about rectangles in triangles. Math. Horiz. **5**, 18–22 (1997)
11. Haurie, A., Marcotte, P.: On the relationship between Nash-Cournot and Wardrop equilibria. Networks **15**, 295–308 (1985)
12. Johari, R., Tsitsiklis, J.N.: Efficiency loss in a network resource allocation game. Math. Oper. Res. **29**, 407–435 (2004)
13. Koutsoupias, E., Papadimitriou, C.H.: Worst-case equilibria. In: Meinel, C., Tison, S. (eds.) Proceedings of the 16th Annual Symposium on Theoretical Aspects of Computer Science. Lecture Notes in Computer Science, Vol. 1563, pp. 404–413, Springer (1999)
14. Lange, L.H.: What is the biggest rectangle you can put inside a given triangle? Coll. Math. J. **24**, 237–240 (1993)
15. Nash, J.: Equilibrium points in n-person games. Proc. Natl. Acad. Sci. **36**, 48–49 (1950)
16. Papadimitriou, C.H.: Algorithms, games, and the Internet. In: Proceedings of the 33rd Annual ACM Symposium on Theory of Computing, pp. 749–753. ACM Press (2001)
17. Pigou, A.C.: The Economics of Welfare. Macmillan (1920)
18. Roughgarden, T.: Potential functions and the inefficiency of equilibria. Proc. Intern. Congr. Math. **III**, 1071–1094 (2006)
19. Roughgarden, T., Tardos, É.: How bad is selfish routing? J. ACM **49**, 236–259 (2002)
20. Roughgarden, T., Tardos, É.: Introduction to the inefficiency of equilibria. Chapter 17 In: Nisan, N. et al. (eds.) Algorithmic Game Theory. Cambridge University Press (2007)
21. Smith, M.J.: The existence, uniqueness and stability of traffic equilibria. Trans. Res. B **13**, 295–304 (1979)
22. Wardrop, J.G.: Some theoretical aspects of road traffic research. Proc. Inst. Civil Eng. Part II **1**, 325–378 (1952)

Resource Buying Games

Tobias Harks and Britta Peis

Abstract In *resource buying games* a set of players jointly buys a subset of a finite resource set E (e.g., machines, edges, or nodes in a digraph). The cost of a resource e depends on the number (or load) of players using e, and has to be paid completely by the players before it becomes available. Each player i needs at least one set of a predefined family $\mathscr{S}_i \subseteq 2^E$. Thus, resource buying games can be seen as a variant of congestion games in which the load-dependent costs of the resources can be shared arbitrarily among the players. A strategy of player i in resource buying games is a tuple consisting of one of i's desired configurations $S_i \in \mathscr{S}_i$ together with a payment vector $p_i \in \mathbb{R}_+^E$ indicating how much i is willing to contribute towards the purchase of the chosen resources. In this chapter, we study the existence of pure Nash equilibria (PNE, for short) of resource buying games. In contrast to classical congestion games for which equilibria are guaranteed to exist, the existence of equilibria in resource buying games strongly depends on the underlying structure of the families $\mathscr{S}_i$ and the behavior of the cost functions. We show that for marginally non-increasing cost functions, matroids are the right structure to consider.

1 Introduction

We introduce and study *resource buying games* as a means to model selfish behavior of players jointly designing a resource infrastructure. In a resource buying game, we are given a finite set N of players and a finite set E of resources. We do not specify the type of the resources, they can be just anything, e.g., edges or nodes in a graph, processors, trucks, etc. In our model, the players jointly buy a subset of the resources. Each player $i \in N$ has a certain weight (demand) d_i, as well as a predefined family

T. Harks (✉)
Institut für Mathematik, Universität Augsburg, Universitätsstraße 14, 86135 Augsburg, Germany
e-mail: tobias.harks@math.uni-augsburg.de

B. Peis
Fakultät für Wirtschaftswissenschaften, RWTH Aachen, Kackertstraße 7,
52072 Aachen, Germany
e-mail: britta.peis@oms.rwth-aachen.de

of subsets (called *configurations*) $\mathscr{S}_i \subseteq 2^E$ from which player i needs at least one configuration $S_i \in \mathscr{S}_i$ to be bought by the players. For example, the families $\mathscr{S}_i$ could be the collection of all paths linking two player-specific terminal-nodes s_i, t_i in a digraph $G = (V, E)$ (such games are called *connection games*), or $\mathscr{S}_i$ could stand for the set of machines on which i can process her job (so-called *scheduling games*). The cost c_e of a resource $e \in E$ depends on the load, i.e., the total weight of players using e. This load-dependent cost needs to be paid completely by the players before the resource becomes available. As usual, we assume that the cost functions c_e are non-decreasing and normalized in the sense that c_e never decreases with increasing load, and that c_e is zero if none of the players is using e. In resource buying games, a strategy of player i is a tuple (S_i, p_i) consisting of one of i's desired sets $S_i \in \mathscr{S}_i$, together with a payment vector $p_i \in \mathbb{R}_+^E$ indicating how much i is willing to contribute to the purchase of the resources. The goal of each player is to pay as little as possible while ensuring that the bought resources contain at least one of her desired configurations. A *pure strategy Nash equilibrium* (PNE, for short) is a strategy profile $\{(S_i, p_i)\}_{i \in N}$ such that none of the players has an incentive to switch her strategy given that the remaining players stick to the chosen strategy. A formal definition of the model will be given in Sect. 2.

First Insights

We start with two examples.

Example 1 Consider the scheduling game illustrated in Fig. 1a with two resources (machines) $\{e, f\}$ and three players $\{1, 2, 3\}$ each having unit-sized jobs. Any job fits on any machine, and the processing cost of machines e and f is given by $c_j(\ell_j(S))$, where $\ell_j(S)$ denotes the number of jobs on machine $j \in \{e, f\}$ under schedule S. In our model, each player chooses a strategy, which is a tuple consisting of one of the two machines, together with a payment vector indicating how much she is willing to pay for each of the machines. Now, suppose the cost functions for the two machines are $c_e(0) = c_f(0) = 0$, $c_e(1) = c_f(1) = 1$, $c_e(2) = c_f(2) = 1$ and $c_e(3) = c_f(3) = M$ for some large $M > 0$. One can easily verify that there is no PNE: If two players share the cost of one machine, then a player with positive payments deviates to the other machine. By the choice of M, the case that all players share a single machine can never be a PNE.

In light of this quite basic example, we have to restrict the set of feasible cost functions. Although the cost functions c_e and c_f of the machines in this scheduling

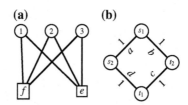

Fig. 1 Scheduling game (**a**) and connection game (**b**)

game are monotonically non-decreasing, their marginal cost function is neither non-increasing, nor non-decreasing, where we call cost function $c_e : \mathbb{N} \to \mathbb{R}_+$ *marginally non-increasing* if

$$c_e(x+\delta) - c_e(x) \geq c_e(y+\delta) - c_e(y) \quad \forall x \leq y;\ x, y, \delta \in \mathbb{N}.$$

We call a cost function c_e marginally non-decreasing if $-c_e$ is marginally non-increasing. Note that marginally non-increasing cost functions model economies of scale and include fixed costs as a special case. They can be seen as a discrete analog of concave functions. Now, consider a scheduling game with unit-sized jobs and marginally non-increasing cost functions. It is not hard to establish a simple polynomial-time algorithm to compute a PNE for this setting: Sort the machines with respect to the costs evaluated at load one. Iteratively, let the player whose minimal cost among her available resources is maximal exclusively pay for that resource, drop this player from the list and update the cost on the bought resource with respect to a unit increment of load.

While the above algorithm might give hope for obtaining a more general existence and computability result for PNE for games with non-increasing marginal cost functions, the following example shows that already very simple connection games do not necessarily admit a PNE:

Example 2 Consider the connection game illustrated in Fig. 1b, where there are two players of unit demand that want to establish an s_i-t_i path for $i = 1, 2$. Any strategy profile (*state*) of the game contains two paths, one for each player, that have exactly one edge e in common. In a PNE, no player would ever pay a positive amount for an edge that is not on her chosen path. Now, a player paying a positive amount for e (and at least one such player exists) would have an incentive to switch strategies as she could use the edge that is exclusively used (and paid) by the other player for free. Note that this example uses fixed costs, which are marginally non-increasing.

Outline

We study resource buying games and investigate the existence of pure-strategy Nash equilibria. In light of the examples illustrated in Fig. 1, we find that equilibrium existence is strongly related to two key properties of the game: the monotonicity of the marginal cost functions and the combinatorial structure of the allowed strategy spaces of the players.

In this paper, we restrict to non-increasing marginal cost functions and investigate the combinatorial structure of the strategy spaces of the players for which PNE exist. It turns out that *matroids* are the right structure to consider in this setting. Matroids have a rich combinatorial structure and include, for instance, scheduling games, as well as the setting where each player wants to build a spanning tree in a graph.

After a more precise definition of resource buying games in Sect. 2, we prove in Sect. 3 that for non-increasing marginal costs and matroid structure, every socially optimal configuration profile can be obtained as a PNE. The proof relies on a complete characterization of the configuration profiles that can appear as a PNE.

2 The Model

A *congestion model* is a tuple $\mathcal{M} = (N, E, \mathcal{S}, (d_i)_{i \in N}, (c_e)_{e \in E})$ consisting of a player set $N = \{1, \ldots, n\}$, a resource set $E = \{1, \ldots, m\}$, and a set of states $\mathcal{S} = \times_{i \in N} \mathcal{S}_i$, also called *configuration profiles*. For each player $i \in N$, the set $\mathcal{S}_i$ is a non-empty set of subsets $S_i \subseteq E$, called *the configurations of i*.

Given a state $S \in \mathcal{S}$, we define $\ell_e(S) = \sum_{i \in N : e \in S_i} d_i$ as the total load of e in S. Every resource $e \in E$ has a *cost function* $c_e : \mathcal{S} \to \mathbb{N}$ defined as $c_e(S) = c_e(\ell_e(S))$. We call a configuration profile $S \in \mathcal{S}$ *(socially) optimal* if its total cost $c(S) = \sum_{e \in E} c_e(S)$ is minimal among all $S \in \mathcal{S}$.

The *resource buying game* associated to congestion model $\mathcal{M}$ is the strategic game on player set N in which each player $i \in N$ chooses a configuration $S_i \in \mathcal{S}_i$ together with a payment vector $p_i \in \mathbb{R}_+^E$ as strategy. Given a strategy profile (S, p) with $S = (S_1, \ldots, S_n) \in \mathcal{S}$ and $p = (p_1, \ldots, p_n) \in \mathbb{R}_+^{N \times E}$, we say that a resource $e \in E$ is *bought* under (S, p), if $\sum_{i \in N} p_i^e \geq c_e(\ell_e(S))$, where p_i^e denotes the payment of player i for resource e. The private cost function of each player $i \in N$ is defined as $\pi_i(S) = \sum_{e \in E} p_i^e$ if S_i is bought, and $\pi_i(S) = \infty$, otherwise. The special case where $d_i = 1$ for all $i \in N$ is called *unweighted* game. We are interested in the existence of pure Nash equilibria, i.e., strategy profiles that are resilient against unilateral deviations. Formally, a strategy profile (S, p) is a PNE, if $\pi_i(S, p) \leq \pi_i((S_i', S_{-i}), (p_i', p_{-i}))$ for all players $i \in N$ and all strategies $(S_i', p_i') \in \mathcal{S}_i \times \mathbb{R}_+^E$. Note that for PNE, we may assume w.l.o.g. that a pure strategy (S_i, p_i) of player i satisfies $p_i^e \geq 0$ for all $e \in S_i$ and $p_i^e = 0$, else.

Definition 3 (*Matroid games*) We call a resource buying game associated to congestion model $\mathcal{M} = (N, E, \mathcal{S}, (d_i)_{i \in N}, (c_e)_{e \in E})$ a *matroid (resource buying) game* if $E = \bigcup_{i \in N} E_i$, and each configuration set $\mathcal{S}_i \subseteq 2^{E_i}$ forms the base set of some matroid $\mathcal{M}_i = (E_i, \mathcal{S}_i)$.

As usual in matroid theory, we will write $\mathcal{B}_i$ instead of $\mathcal{S}_i$, and $\mathcal{B}$ instead of $\mathcal{S}$, when considering matroid games. Recall that a non-empty anti-chain[1] $\mathcal{B}_i \subseteq 2^{E_i}$ is the base set of a matroid $\mathcal{M}_i = (E_i, \mathcal{B}_i)$ on resource (*ground*) set E_i if and only if the following *basis exchange property* is satisfied: whenever $X, Y \in \mathcal{B}_i$ and $x \in X \setminus Y$, then there exists some $y \in Y \setminus X$ such that $X \setminus \{x\} \cup \{y\} \in \mathcal{B}_i$.

It is not hard to see that the scheduling and spanning tree games mentioned above belong to the class of matroid games. Matroids have a very rich combinatorial structure that allows for various alternative characterizations. One well-known matroid property which turns out to be useful for proving the existence of PNE in matroid games below is the following: Given a matroid $M = (E, \mathcal{B})$ with weight function $w : E \to \mathbb{R}_+$, a basis $B \in \mathcal{B}$ is a minimum weight basis of M if and only if there exists no basis B^* with $|B \setminus B^*| = 1$ and $w(B^*) < w(B)$.

---

[1] $\mathcal{B}_i \subseteq 2^{E_i}$ is an *anti-chain* (w.r.t. $(2^{E_i}, \subseteq)$) if $B, B' \in \mathcal{B}_i$, $B \subseteq B'$ implies $B = B'$.

Since a strategy profile (B, p) of a matroid game with $B = (B_1, \ldots, B_n) \in \mathscr{B}$ is a PNE if none of the players $i \in N$ can improve by switching to some other basis $B'_i \in \mathscr{B}_i$, given that all other players $j \neq i$ stick to their chosen strategy (B_j, p_j), it suffices to consider bases $\hat{B}_i \in \mathscr{B}_i$ with $\hat{B}_i = B_i - e + f$ for some $e \in B_i \setminus \hat{B}_i$ and $f \in \hat{B}_i \setminus B_i$. Note that by switching from B_i to $\hat{B}_i$, player i would need to pay the additional marginal cost $c_f(\ell_f(B) + d_i) - c_f(\ell_f(B))$, but would not need to pay for element e. We get the following proposition.

Proposition 4 *A strategy profile (B, p) of a matroid game is a PNE if and only if*

$$p_i^e \leq c_f(\ell_f(B) + d_i) - c_f(\ell_f(B))$$

for all $i \in N$ and all $\hat{B}_i \in \mathscr{B}_i$ with $\hat{B}_i = B_i - e + f$ for some $e \in B_i \setminus \hat{B}_i$ and $f \in \hat{B}_i \setminus B_i$.

3 Existence of PNE in Matroid Games

Given a matroid game associated to congestion model $(N, E, \mathscr{B}, d, c)$, we call a configuration profile $B = (B_1, \ldots, B_n) \in \mathscr{B}$ *implementable as PNE* if there exists a payment vector $p = (p_1, \ldots, p_n) \in \mathbb{R}_+^{N \times E}$ such that profile (B, p) is a PNE.

In Theorem 5 below, we derive a characterization of configuration profiles that are implementable as PNE. Afterwards, in Theorem 6, we show that each socially optimal configuration profile $B \in \mathscr{B}$ of a matroid game with non-increasing marginal costs satisfies the condition of Theorem 5. As a consequence, each matroid game with non-increasing marginal costs admits a PNE, even a socially optimal one.

To state the two theorems, we need a couple of definitions: For $B \in \mathscr{B}, e \in E$ and $i \in N_e(B) := \{i \in N \mid e \in B_i\}$, let $\text{ex}_i(e) := \{f \in E - e \mid B_i - e + f \in \mathscr{B}_i\} \subseteq E$ denote the set of all resources f such that player i could exchange the resources e and f to obtain an alternative basis $B_i - e + f \in \mathscr{B}_i$. Note that $\text{ex}_i(e)$ might be empty, and that, if $\text{ex}_i(e)$ is empty, the element e lies in every basis of player i (by the matroid basis exchange property). Let $F := \{e \in E \mid e \text{ lies in each basis of } i \text{ for some } i \in N\}$ denote the set of elements that are fixed in the sense that they must lie in one of the players' chosen bases. Furthermore, we define for all $e \in E - F$ and all $i \in N_e(B)$ and all $f \in \text{ex}_i(e)$ the value $\Delta_i(B; e \to f) := c_f(\ell_f(B_i + f - e, B_{-i})) - c_f(\ell_f(B))$ which is the marginal amount that needs to be paid in order to buy resource f if i switches from B_i to $B_i - e + f$. Finally, for all $e \in E - F$ and all $i \in N_e(B)$, define

$$\Delta_i^e(B) := \min_{f \in \text{ex}_i(e)} \{\Delta_i(B; e \to f)\}. \tag{1}$$

Theorem 5 *Given a matroid game associated to congestion model $(N, E, \mathscr{B}, d, c)$, a configuration profile $B \in \mathscr{B}$ is implementable as PNE if and only if*

$$c_e(B) \leq \sum_{i \in N_e(B)} \Delta_i^e(B) \quad \text{for all } e \in E \setminus F. \tag{2}$$

Proof We first proof necessity. Let (B, p) be a PNE. Then, by Proposition 4 and the definition of a PNE, we obtain for all $e \in E \setminus F$:

$$c_e(B) = \sum_{i \in N_e(B)} p_i^e \leq \sum_{i \in N_e(B)} \Delta_i^e(B).$$

Note that the $\Delta_i^e(B)$-values are well defined as we only consider elements in $E \setminus F$.

Now, we prove sufficiency. For all $e \in F$ we pick a player i with $\text{ex}_i(e) = \emptyset$ and let her pay the entire cost, i.e., $p_i^e = c_e(B)$. For all $e \in E \setminus F$ and $i \in N_e(B)$, we define

$$p_i^e = \frac{\Delta_i^e(B)}{\sum_{j \in N_e(B)} \Delta_j^e(B)} \cdot c_e(B),$$

if the denominator is positive, and $p_i^e = 0$, otherwise. Using (2), we obtain $p_i^e \leq \Delta_i^e(B)$ for all $e \in E \setminus F$, proving that (B, p) is a PNE. $\square$

Note that the above characterization holds for arbitrary non-negative and non-decreasing cost functions. In particular, if property (2) were true for a given configuration profile $B \in \mathcal{B}$, it follows from the constructive proof that the payment vector p can be computed efficiently. The following Theorem 6 states that matroid games with non-increasing marginal costs always possess a PNE. We prove Theorem 6 by showing that any socially optimal configuration $B \in \mathcal{B}$ satisfies (2).

Theorem 6 *Every matroid game with marginally non-increasing cost functions possesses a PNE.*

Proof We prove that any socially optimal configuration profile $B \in \mathcal{B}$ satisfies (2) and, thus, by Theorem 5 there exists a payment vector p such that (B, p) is a PNE. Assume by contradiction that B does not satisfy (2). Hence, there is an $e \in E \setminus F$ with

$$c_e(B) > \sum_{i \in N_e(B)} \Delta_i^e(B). \tag{3}$$

By relabeling indices we may write $N_e(B) = \{1, \ldots, k\}$ for some $1 \leq k \leq n$, and define for every $i \in N_e(B)$ the tuple $(\hat{B}_i, f_i) \in \mathcal{B}_i \times (E_i - e)$ such that $\hat{B}_i = B_i + f_i - e$ and $f_i \in \arg\min_{f \in \text{ex}_i(e)} \{\Delta_i(B; e \to f)\}$. Thus, $\Delta_i^e(B) = \Delta_i(B; e \to f_i)$. Note that $(\hat{B}_i, f_i)$ is well defined as $e \in E \setminus F$. We now iteratively change the current basis of every player in $N_e(B)$ in the order of their indices to the alternative basis $\hat{B}_i, i = 1, \ldots, k$. This gives a sequence of profiles $(B^0, B^1, \ldots, B^k)$ with $B^0 = B$ and $B^i = (\hat{B}_i, B_{-i}^{i-1})$ for $i = 1, \ldots, k$. For the cost increase of the new elements $f_i, i \in N_e(B)$, we obtain the key inequality

$$c_{f_i}(\ell_{f_i}(B^i)) - c_{f_i}(\ell_{f_i}(B^{i-1})) \leq \Delta_i^e(B). \tag{4}$$

This inequality holds because for elements f_i with $f_i \neq f_j$ for all $j \in N_e(B), j < i$, we have by (1) $c_{f_i}(\ell_{f_i}(B^i)) - c_{f_i}(\ell_{f_i}(B^{i-1})) = \Delta_i^e(B)$. And for elements f_i with $f_i = f_j$ for possibly several $j \in N_e(B), j < i$, we have $c_{f_i}(\ell_{f_i}(B^i)) - c_{f_i}(\ell_{f_i}(B^{i-1})) \leq \Delta_i^e(B)$ as cost functions are marginally non-increasing. Putting everything together, yields

$$\begin{aligned}
c(B) - c(B^k) &= \sum_{i=1}^{k}(c(B^{i-1}) - c(B^i)) \\
&= \sum_{i=1}^{k}(c_e(\ell_e(B^{i-1})) + c_{f_i}(\ell_{f_i}(B^{i-1})) - c_e(\ell_e(B^i)) - c_{f_i}(\ell_{f_i}(B^i))) \\
&= c_e(\ell_e(B)) - c_e(\ell_e(B^k)) - \sum_{i=1}^{k}(c_{f_i}(\ell_{f_i}(B^i)) - c_{f_i}(\ell_{f_i}(B^{i-1}))) \\
&\geq c_e(\ell_e(B)) - \sum_{i=1}^{k}\Delta_i^e(B) \\
&> 0,
\end{aligned}$$

where the first inequality uses (4) and $c_e(\ell_e(B^k)) = c_e(0) = 0$ (note that $e \in E \setminus F$). The second, strict inequality follows from (3). Altogether, we obtain a contradiction to the optimality of B. □

Bibliographic Notes

In the first paper in the area of resource buying games, Anshelevich et al. [5] introduced *connection games* to model selfish behavior of players jointly designing a network infrastructure. Their model can be seen as the special class of resource buying games in which player i's configuration set consists of all $s_i \to t_i$-paths linking two player-specific terminal nodes s_i and t_i in a commonly used undirected graph $G = (V, E)$, and the costs $c_e, e \in E$, are fixed. Anshelevich et al. showed that these games have a PNE if all players connect to a common source. They also show that general connection games might fail to have a PNE, see Example 2 above. Several follow-up papers (cf. [2–4, 6–8, 11, 13]) study the existence and efficiency of pure Nash and strong equilibria in connection games and extensions of them.

Hoefer [12] studied resource buying games for load-dependent non-increasing marginal cost functions, generalizing fixed costs. He considers unweighted congestion games modeling covering and facility location problems. Among other results regarding approximate PNEs and the price of anarchy/stability, he gives a polynomial-time algorithm computing a PNE for the special case where every player wants to cover a single element.

This chapter is based on our paper [10], in which we show that the statement of Theorem 6 is tight in the following sense: For every two-player resource buying game with non-matroidal strategy configurations, one can construct an isomorphic game with non-increasing marginal costs that does not admit a PNE. Here, the term "isomorphic" means that in the constructed game, we do not take into account how the non-matroid strategy spaces of the two players actually interweave (cf. Ackermann et al. [1] who introduced the notion of interweaving of strategy spaces). Moreover, we present a polynomial-time algorithm to compute a PNE for unweighted matroid resource buying games with marginally non-increasing cost functions. This algorithm can be regarded as a non-trivial extension of the simple algorithm for scheduling games described before: starting with the collection of matroids, our algorithm iteratively makes use of deletion and contraction operations, until a basis together with a suitable payment vector for each of the players is found.

Furthermore, we show that every resource buying game with *non-decreasing marginal costs* possesses a PNE regardless of the strategy space. We prove this result by showing that an optimal configuration profile can be obtained as a PNE. We further show that for marginally increasing costs one can compute a PNE efficiently whenever one can compute a best response efficiently.

The characterization stated in Theorem 5 is inspired by a similar result given in von Falkenhausen and Harks [9], where enforceable strategy profiles (via cost shares) are characterized.

References

1. Ackermann, H., Röglin, H., Vöcking, B.: Pure Nash equilibria in player-specific and weighted congestion games. Theor. Comput. Sci. **410**(17), 1552–1563 (2009)
2. Anshelevich, E., Caskurlu, B.: Exact and approximate equilibria for optimal group network formation. Theor. Comput. Sci. **412**(39), 5298–5314 (2011)
3. Anshelevich, E., Caskurlu, B.: Price of stability in survivable network design. Theory Comput. Syst. **49**(1), 98–138 (2011)
4. Anshelevich, E., Caskurlu, B., Hate, A.: Strategic multiway cut and multicut games. Theory Comput. Syst. **52**(2), 200–220 (2013)
5. Anshelevich, E., Dasgupta, A., Tardos, É., Wexler, T.: Near-optimal network design with selfish agents. Theory Comput. **4**(1), 77–109 (2008)
6. Anshelevich, E., Karagiozova, A.: Terminal backup, 3D matching, and covering cubic graphs. SIAM J. Comput. **40**(3), 678–708 (2011)
7. Cardinal, J., Hoefer, M.: Non-cooperative facility location and covering games. Theor. Comput. Sci. **411**, 1855–1876 (2010)
8. Epstein, A., Feldman, M., Mansour, Y.: Strong equilibrium in cost sharing connection games. Games Econ. Behav. **67**(1), 51–68 (2009)
9. von Falkenhausen, P., Harks, T.: Optimal cost sharing for resource selection games. Math. Oper. Res. **38**(1), 184–208 (2013)
10. Harks, T., Peis, B.: Resource buying games. Algorithmica **70**(3), 493–512 (2014)
11. Hoefer, M.: Non-cooperative tree creation. Algorithmica **53**, 104–131 (2009)
12. Hoefer, M.: Strategic cooperation in cost sharing games. Int. J. Game Theory **42**(1), 29–53 (2013)

13. Hoefer, M., Skopalik, A.: On the complexity of Pareto-optimal Nash and strong equilibria. In: Konogiannis, S., Koutsoupias, E., Spirakis, P. (eds.) Proceedings of the 3rd International Symposium Algorithmic Game Theory. Lecture Notes in Computer Science, vol. 6386, pp. 312–322 (2010)

Linear, Exponential, but Nothing Else

On Pure Nash Equilibria in Congestion Games and Priority Rules for Single-Machine Scheduling

Max Klimm

Abstract We consider two seemingly unrelated resource allocation problems and show that they share a deep structural property. In the first problem, we schedule jobs on a single machine to minimize the sum of the jobs' cost where each job's cost is determined by a job-specific function of its completion time. In the second problem, we consider weighted congestion games and are interested in the existence of pure Nash equilibria. We show that the classes of delay cost functions for which the scheduling problem admits a priority rule are exactly the classes of resource cost functions that guarantee the existence of a pure Nash equilibrium in weighted congestion games. These classes of cost functions are those that contain only linear functions or exponential functions, but not both.

1 Problem Description

1.1 Single-Machine Scheduling

We are given a set $N = \{1, \ldots, n\}$ of n jobs with processing times $p_j > 0$ and delay cost functions $g_j : \mathbb{R}_{\geq 0} \to \mathbb{R}$ to be scheduled on a single machine. Since the machine may not process two or more jobs at the same time, a feasible schedule has to decide on a sequence in which the jobs are processed one after another. The completion time of job j in a particular schedule S is denoted by C_j and the goal is to minimize the total delay costs $G(S) = \sum_{j \in N} g_j(C_j)$.

An interesting class of algorithms to solve these problems is the class of priority rules. A priority rule assigns to each job j a priority index $\delta_j \in \mathbb{R}$ which may depend

---

M. Klimm (✉)
Institut für Mathematik, Technische Universität Berlin, Straße des 17. Juni 136, 10623 Berlin, Germany
e-mail: klimm@math.tu-berlin.de

on the job's processing time and delay cost function, and schedules the jobs in non-increasing order of δ_j. Smith's rule states that for linear delay cost functions of the form $g_j(C_j) = a_j C_j$, $a_j > 0$, the ratio $\delta_j := a_j/p_j$ is such an index that always minimizes the total delay costs.

Our aim is to explore the full space of delay cost functions for which such priority indices exist. To this end we call a set $\mathscr{C}$ of functions *consistent* for single-machine scheduling if there is a priority index $\delta_j : \mathbb{R}_{>0} \times \mathscr{C} \to \mathbb{R}$ such that, for each instance with delay cost functions in $\mathscr{C}$, the order given by $\delta_j(p_j, g_j)$ minimizes the total delay costs.

1.2 Weighted Congestion Games

We are given a set $N = \{1, \ldots, n\}$ of n players and a set R of resources. Each player has a demand $d_i > 0$ and a set of strategies $S_i \subseteq 2^R$ where every strategy $s_i \in S_i$ is a subset of the resources. A tuple of strategies $\mathbf{s} = (s_1, \ldots, s_n)$, one for each player, is called a strategy profile. The players strive to minimize their private cost which is defined in terms of cost functions of the resources. For each resource r we are given a cost function $c_r : \mathbb{R}_{\geq 0} \to \mathbb{R}$ and the private cost of player i in strategy profile $\mathbf{s}$ is defined as $\pi_i(\mathbf{s}) = \sum_{r \in s_i} c_r(x_r(\mathbf{s}))$, where $x_r(\mathbf{s}) = \sum_{j \in N : r \in s_j} d_j$.

We are interested in the existence of pure Nash equilibria in these games. A strategy profile $\mathbf{s}$ is a pure Nash equilibrium, if no player can decrease her private cost by a unilateral deviation, i.e., $\pi_i(t_i, \mathbf{s}_{-i}) \geq \pi_i(\mathbf{s})$ for all players i and strategies $t_i \in S_i$, where we denote by $(t_i, \mathbf{s}_{-i})$ the strategy profile in which all players except player i play as in $\mathbf{s}$ and player i plays t_i.

In order to capture which cost functions guarantee the existence of a pure Nash equilibrium we call a set $\mathscr{C}$ of functions *consistent* for weighted congestion games if each weighted congestion game with resource cost functions in $\mathscr{C}$ has a pure Nash equilibrium.

1.3 Chapter Notes

In this chapter, we prove that the classes of delay cost functions for which single-machine scheduling admits a priority rule are exactly the classes of resource cost functions that guarantee the existence of a pure Nash equilibrium in weighted congestion games, and that these classes of functions are those that contain only linear functions or contain only certain exponential functions. While these results are known [7, 18], the focus of this chapter lies on a universal treatment of both problems. To this end, we develop a functional equation that needs to be satisfied both by delay cost functions that are consistent for single-machine scheduling and resource cost functions that are consistent for weighted congestion games. To keep this chapter concise, we make the simplifying assumption that the cost functions are strictly increasing.

This assumption is not necessary to obtain the result, as neither Harks and Klimm [7] nor Rothkopf and Smith [18] require it. Rothkopf and Smith, however, assume that the delay cost functions are twice continuously differentiable (up to isolated points), while, in this chapter, we require only continuity.

2 Sufficient Conditions on Consistency

We first give sufficient conditions on the consistency of cost functions. Specifically, we show that both the set

$$\mathscr{C}_{\text{lin}} := \{f : \mathbb{R}_{\geq 0} \to \mathbb{R}_{\geq 0} \mid f(x) = ax + b \text{ with } a, b \in \mathbb{R}\}$$

of linear functions and the set

$$\mathscr{C}_{\text{exp}}(\phi) := \{f : \mathbb{R}_{\geq 0} \to \mathbb{R}_{\geq 0} \mid f(x) = ae^{\phi x} + b \text{ with } a, b \in \mathbb{R}\}$$

of exponential functions with joint coefficient ϕ are consistent for both single-machine scheduling and weighted congestion games.

We start with the result for single-machine scheduling.

Theorem 1 *Let $\mathscr{C}$ be a set of functions. If $\mathscr{C} \subseteq \mathscr{C}_{\text{lin}}$ or $\mathscr{C} \subseteq \mathscr{C}_{\text{exp}}(\phi)$ with $\phi \in \mathbb{R}$, then $\mathscr{C}$ is consistent for single-machine scheduling.*

Proof As constant terms do not matter in the individual jobs' cost functions we assume that either $g_j(x) = a_j x$ with $a_j \in \mathbb{R}$ for all jobs j, or there is $\phi \neq 0$ such that $g_j(x) = a_j e^{\phi x}$ with $a_j \in \mathbb{R}$ for all jobs j. Consider the following priority indices:

$$\delta_j = \frac{a_j}{p_j}, \text{ if } g_j(x) = a_j x, \qquad \delta_j = \frac{a_j e^{\phi p_j}}{e^{\phi p_j} - 1}, \text{ if } g_j(x) = a_j e^{\phi x}.$$

The optimality of these priority rules follows from an exchange argument: Suppose there is an optimal schedule S that does not order the jobs according to these priority rules. In this schedule, there are at least two consecutive jobs i, j such that i precedes j although $\delta_i < \delta_j$. Let t denote the total processing times of the jobs before i in S and consider the schedule S' that is obtained from S by swapping jobs i and j. For the linear case, we calculate

$$G(S') - G(S) = a_i p_j - a_j p_i = p_i p_j \left(\frac{a_i}{p_i} - \frac{a_j}{p_j}\right) = p_i p_j (\delta_i - \delta_j) < 0.$$

Similarly, we obtain for the exponential case

$$\begin{aligned}G(S') - G(S) &= a_i e^{\phi(t+p_i+p_j)} - a_i e^{\phi(t+p_i)} + a_j e^{\phi(t+p_j)} - a_j e^{\phi(t+p_i+p_j)} \\&= e^{\phi t}\left(a_i e^{\phi p_i}(e^{\phi p_j} - 1) - a_j e^{\phi p_j}(e^{\phi p_i} - 1)\right) \\&= e^{\phi t}(e^{\phi p_i} - 1)(e^{\phi p_j} - 1)\left(\frac{a_i e^{\phi p_i}}{e^{\phi p_i} - 1} - \frac{a_j e^{\phi p_j}}{e^{\phi p_j} - 1}\right) \\&= e^{\phi t}(e^{\phi p_i} - 1)(e^{\phi p_j} - 1)(\delta_i - \delta_j) < 0.\end{aligned}$$

In both the linear and exponential case this contradicts the optimality of S. □

Perhaps surprisingly, one can show that the same set of cost functions is consistent for weighted congestion games as well.

Theorem 2 *Let $\mathscr{C}$ be a set of functions. If $\mathscr{C} \subseteq \mathscr{C}_{lin}$ or $\mathscr{C} \subseteq \mathscr{C}_{exp}(\phi)$ with $\phi \in \mathbb{R}$, then $\mathscr{C}$ is consistent for weighted congestion games.*

Proof Fix a congestion game $G = (N, \mathbf{S}, \boldsymbol{\pi})$ with only linear or only exponential functions. We proceed to show that G admits a potential function and, thus, has a pure Nash equilirium. A potential function is a function $P : \mathbf{S} \to \mathbb{R}$ with the property that $P(t_i, \mathbf{s}_{-i}) < P(\mathbf{s})$ for all $\mathbf{s} \in \mathbf{S}$, $i \in N$ and $t_i \in S_i$ with $\pi_i(t_i, \mathbf{s}_{-i}) < \pi_i(\mathbf{s})$. By construction, every local minimum of a potential function is a pure Nash equilibrium.

We are going to show the existence of a potential function of the form

$$P(\mathbf{s}) = \sum_{i \in N} \sum_{r \in s_i} \lambda_i c_r \left(\sum_{j \in \{1,\ldots,i\}: r \in s_j} d_j \right), \quad (1)$$

where $\lambda_i \in \mathbb{R}_{>0}$ is a player-specific scaling factor to be determined later. The value $P(\mathbf{s})$ can be interpreted as the sum of the players' costs scaled by λ_i where each player i pays only resource costs as if players with larger index are not present in the game. In the remainder of the proof, we will choose the scaling factor λ_i such that the potential function is independent of the ordering of the players.

If $\mathscr{C}$ contains only linear functions, we set $\lambda_i = d_i$ for all players i. It is easy to verify that the potential function can be reformulated as

$$P(\mathbf{s}) = \frac{1}{2} \sum_{r \in R} \left(c_r(x_r(\mathbf{s})) + c_r(0)\right) x_r(\mathbf{s}) + \frac{1}{2} \sum_{i \in N} \sum_{r \in s_i} (c_r(d_i) - c_r(0)) d_i,$$

see Fig. 1 for a graphical illustration of this fact. This expression is independent of the ordering of the players.

If $\mathscr{C} \subseteq \mathscr{C}_{exp}(\phi)$ with $\phi \neq 0$, we set $\lambda_i = \mathrm{sgn}(\phi)(1 - e^{-\phi d_i})$ and obtain

$$P(\mathbf{s}) = \mathrm{sgn}(\phi) \sum_{i \in N} \sum_{r \in s_i} (1 - e^{-\phi d_i}) c_r \left(\sum_{j \in \{1,\ldots,i\}: r \in s_j} d_j \right).$$

Linear, Exponential, but Nothing Else

Fig. 1 Potential $d_1 c_r(d_1) + d_2 c_r(d_1 + d_2) + d_3 c_r(d_1 + d_2 + d_3)$ of a resource r with cost function c_r used by players 1, 2, and 3. The potential corresponds to the *colored area* and is independent of the ordering of the players.

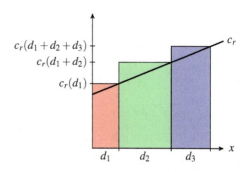

Using that $c_r = a_r e^{\phi x} + b_r$ for all $r \in R$, this gives

$$P(\mathbf{s}) = \mathrm{sgn}(\phi) \sum_{r \in R} \left(a_r \left(e^{\phi x(\mathbf{s})} - 1 \right) + b_r \sum_{i \in N : r \in s_i} \left(1 - e^{-\phi d_i} \right) \right),$$

which does not depend on the ordering of the players either.

Given that the potential function (1) is independent of the ordering of the players, it is without loss of generality to assume that player n deviates. For arbitrary $\mathbf{s} \in \mathbf{S}$, $t_n \in S_n$ with $\pi_n(t_n, \mathbf{s}_{-n}) < \pi_n(\mathbf{s})$, we then obtain

$$P(\mathbf{s}) - P(t_n, \mathbf{s}_{-n}) = \lambda_n \sum_{r \in s_n \setminus t_n} c_r(x(\mathbf{s})) - \lambda_n \sum_{r \in t_n \setminus s_n} c_r(x(t_n, \mathbf{s}_{-n}))$$
$$= \lambda_n \left(\pi_n(\mathbf{s}) - \pi_n(t_n, \mathbf{s}_{-n}) \right) > 0.$$

Thus, P is a potential function and G has a pure Nash equilibrium. □

3 Necessary Conditions on Consistency

We proceed to derive necessary conditions on the consistency of cost functions for single-machine scheduling and weighted congestion games. Throughout this chapter, we assume that $\mathscr{C}$ is closed under positive integer scalar multiplication, i.e., $f \in \mathscr{C}$ implies $vf \in \mathscr{C}$ for all $v \in \mathbb{N}$. This assumption is without loss of generality for weighted congestion games, where a resource with cost function vf, $v \in \mathbb{N}$, can be simulated by v copies of a resource with cost function f. In the context of scheduling games, the scalar v associated with the delay cost function of a job corresponds to the weight of that job. We note that we here require only that $\mathscr{C}$ is closed under positive *integer* scalar multiplication, while Rothkopf and Smith require closure under *arbitrary* positive scalar multiplication for their characterization.

Lemma 3 *Let $\mathscr{C}$ be a set of strictly increasing functions that is closed under positive integer scalar multiplication. If $\mathscr{C}$ is consistent for single-machine scheduling, then*

$$\frac{f(x+y)-f(x)}{h(x+y)-h(y)} = \frac{f(x+y+t)-f(x+t)}{h(x+y+t)-h(y+t)} \qquad (S)$$

for all $f, h \in \mathscr{C}$ and $x, y, t, \in \mathbb{R}_{\geq 0}$.

Proof Consider two jobs i, j with processing times $p_i = x$ and $p_j = y$ and cost functions $g_i = vf$ and $g_j = uh$ for some $u, v \in \mathbb{N}$. For the existence of a priority rule it is necessary that the following invariance condition is satisfied; if

$$vf(x+y) - vf(x) \geq uh(x+y) - uh(y), \qquad (2)$$

i.e., job i should precede job j when both are consecutively scheduled in the beginning, then also

$$vf(x+y+t) - vf(x+t) \geq uh(x+y+t) - uh(y+t) \qquad (3)$$

for all $t \geq 0$, i.e., job i should precede job j when both are consecutively scheduled after time t. If the inequality sign in (2) is reversed, so is the inequality sign in (3). Put differently, for all $u, v \in \mathbb{N}$ at least one of the following two cases holds:

$$\frac{u}{v} \leq \min\left\{\frac{f(x+y)-f(x)}{h(x+y)-h(y)}, \frac{f(x+y+t)-f(x+t)}{h(x+y+t)-h(y+t)}\right\} \quad \text{or} \quad \frac{u}{v} \geq \max\left\{\frac{f(x+y)-f(x)}{h(x+y)-h(y)}, \frac{f(x+y+t)-f(x+t)}{h(x+y+t)-h(y+t)}\right\}.$$

This implies (S), because otherwise for any $u, v \in \mathbb{N}$ with

$$\min\left\{\frac{f(x+y)-f(x)}{h(x+y)-h(y)}, \frac{f(x+y+t)-f(x+t)}{h(x+y+t)-h(y+t)}\right\} < \frac{u}{v} < \max\left\{\frac{f(x+y)-f(x)}{h(x+y)-h(y)}, \frac{f(x+y+t)-f(x+t)}{h(x+y+t)-h(y+t)}\right\}$$

the invariance condition is violated. □

We proceed to give a structurally similar condition on the consistency for weighted congestion games.

Lemma 4 *Let $\mathscr{C}$ be a set of strictly increasing functions that is closed under positive integer scalar multiplication. If $\mathscr{C}$ is consistent for weighted congestion games, then*

$$\frac{f(x+y)-f(x)}{f(x+y)-f(y)} = \frac{h(x+y+t)-h(x+t)}{h(x+y+t)-h(y+t)} \qquad (G)$$

for all $f, h \in \mathscr{C}$ and $x, y, t \in \mathbb{R}_{\geq 0}$.

Proof Let $u, v \in \mathbb{N}$ be arbitrary and consider a weighted congestion game with four resources $R = \{r_1, r_2, r_3, r_4\}$ with cost functions $c_1 = c_4 = uh$, $c_2 = c_3 = vf$ and four players i, j, k, k' with demands $d_i = x$, $d_j = y$, $d_k = d_{k'} = t$ and strategy sets $S_i = \{\{r_1, r_2\}, \{r_3, r_4\}\}$, $S_j = \{\{r_1, r_3\}, \{r_2, r_4\}\}$, $S_k = \{\{r_1\}\}$, $S_{k'} = \{\{r_4\}\}$.

Players i and j have two strategies each and players k and k' have one strategy each resulting in four distinct strategy profiles. There are only two types of strategy profiles: in the first type players i and j are together on a resource with cost function uh, but are not together on a resource with cost function vf. For the second type, it is the other way around. In a strategy profile $\mathbf{s}^1$ of the first type we have

$$\pi_i(\mathbf{s}^1) = uh(x+y+t) + vf(x), \quad \pi_j(\mathbf{s}^1) = uh(x+y+t) + vf(y) \quad (4a)$$

while in a strategy profile $\mathbf{s}^2$ of the second type we have

$$\pi_i(\mathbf{s}^2) = uh(x+t) + vf(x+y), \quad \pi_j(\mathbf{s}^2) = uh(y+t) + vf(x+y). \quad (4b)$$

In order to have a pure Nash equilibrium, one of these two types must be beneficial to both players, i.e., at least one of the following two cases holds:

$$\pi_\ell(\mathbf{s}^1) \leq \pi_\ell(\mathbf{s}^2) \text{ for } \ell \in \{i, j\} \quad \text{or} \quad \pi_\ell(\mathbf{s}^1) \geq \pi_\ell(\mathbf{s}^2) \text{ for } \ell \in \{i, j\}. \quad (5)$$

Combining (4) and (5), we obtain that at least one of the following cases holds:

$$\frac{u}{v} \leq \min\left\{\frac{f(x+y)-f(x)}{h(x+y+t)-h(x+t)}, \frac{f(x+y)-f(y)}{h(x+y+t)-h(y+t)}\right\} \quad \text{or} \quad \frac{u}{v} \geq \max\left\{\frac{f(x+y)-f(x)}{h(x+y+t)-h(x+t)}, \frac{f(x+y)-f(y)}{h(x+y+t)-h(y+t)}\right\}.$$

Analogously to the proof of Lemma 3 we obtain $\frac{f(x+y)-f(x)}{h(x+y+t)-h(x+t)} = \frac{f(x+y)-f(y)}{h(x+y+t)-h(y+t)}$. Multiplying this equation with $\frac{h(x+y+t)-h(x+t)}{f(x+y)-f(y)}$ gives the claimed result. $\square$

4 Solving the Functional Equation

We are now in position to solve the functional equations (S) and (G) that are necessary for the consistency for single-machine scheduling and weighted congestion games, respectively. In fact, we first analyze a simpler functional equation which appears as a joint special case of (S) and (G) when choosing $f = h$.

Corollary 5 *Let $\mathscr{C}$ be a set of strictly increasing functions that is closed under positive integer scalar multiplication. If $\mathscr{C}$ is consistent for single-machine scheduling or consistent for weighted congestion games, then*

$$\frac{f(x+y)-f(x)}{f(x+y)-f(y)} = \frac{f(x+y+t)-f(x+t)}{f(x+y+t)-f(y+t)} \quad (\star)$$

for all $f \in \mathscr{C}$ and all $x, y, t, \in \mathbb{R}_{\geq 0}$.

We proceed to solve the functional equation $(\star)$.

Theorem 6 *Let f be a continuous and strictly increasing function satisfying the functional equation ($\star$) for all $x, y, t \in \mathbb{R}_{\geq 0}$. Then, f is one of the two functional forms:*

$$f(x) = ax + b, \quad \text{or} \quad f(x) = ae^{\phi x} + b.$$

Proof Let $\varepsilon > 0$ be arbitrary and let $x = \varepsilon$, $y = 2\varepsilon$, and $t = m\varepsilon$ for some $m \in \mathbb{N}$. Using ($\star$), we obtain

$$\gamma := \frac{f(3\varepsilon) - f(1\varepsilon)}{f(3\varepsilon) - f(2\varepsilon)} = \frac{f((m+3)\varepsilon) - f((m+1)\varepsilon)}{f((m+3)\varepsilon) - f((m+2)\varepsilon)}$$

for all $m \in \mathbb{N}$. We note that $\gamma \neq 1$ since $f(1\varepsilon) \neq f(2\varepsilon)$. Setting $a_m := f(m\varepsilon)$ for all $m \in \mathbb{N}$ and rearranging terms, we derive that the sequence $(a_m)_{m \in \mathbb{N}}$ obeys the recursive formula

$$0 = a_{m+3} - \frac{\gamma}{\gamma - 1} a_{m+2} + \frac{1}{\gamma - 1} a_{m+1}$$

for all $m \in \mathbb{N}$. The characteristic equation of this recurrence relation equals

$$x^2 - \frac{\gamma}{\gamma - 1} x + \frac{1}{\gamma - 1} = (x - 1)\left(x - \frac{1}{\gamma - 1}\right).$$

If $\gamma \neq 2$, the characteristic equation has two distinct roots and a_m can be calculated explicitly and uniquely as

$$a_m = a \left(\frac{1}{\gamma - 1}\right)^m + b \tag{6}$$

for some constants $a, b \in \mathbb{R}$. If, on the other hand, $\gamma = 2$, we can calculate a_m as

$$a_m = am + b \tag{7}$$

for some constants $a, b \in \mathbb{R}$. Thus, f is affine or exponential for every integer multiple of ε. As ε was arbitrary we may conclude that f is affine or exponential on a dense subset of $\mathbb{R}_{\geq 0}$. Using that f is continuous, the claimed result follows. $\square$

We proceed to prove the main result of this chapter.

Theorem 7 *Let $\mathscr{C}$ be a set of strictly increasing and continuous functions that is closed under positive integer multiplication. Then, the following are equivalent:*

1. *$\mathscr{C}$ is consistent for single-machine scheduling.*
2. *$\mathscr{C}$ is consistent for weighted congestion games.*

3. One of the following two cases holds:
   a. $\mathscr{C}$ contains only linear functions of the form $f(x) = ax + b$.
   b. $\mathscr{C}$ contains only exponential functions of the form $f(x) = ae^{\phi x} + b$, where ϕ is equal for all functions in $\mathscr{C}$.

Proof The directions $3 \Rightarrow 1$ and $3 \Rightarrow 2$ follow from Theorems 1 and 2, respectively.

To see also the implications $1 \Rightarrow 3$ and $2 \Rightarrow 3$ recall that, by Theorem 6, all functions in $\mathscr{C}$ are either linear or exponential. We proceed to show that the conditions (S) and (G) are satisfied only if $\mathscr{C}$ is as claimed. To this end, we have to rule out the possibility that either $\mathscr{C}$ contains both affine and exponential functions or $\mathscr{C}$ contains exponential functions with different ϕ in the exponent.

As for (S), note that, for all $x, y, t \in \mathbb{R}_{\geq 0}$, the ratio $\frac{f(x+y)-f(x)}{f(x+y+t)-f(x+t)}$ is equal to 1, if f is linear and is equal to $e^{-\phi t}$ if f is of the form $ae^{\phi x} + b$. Thus, (S) is only satisfied for all $t \in \mathbb{R}_{\geq 0}$ if $\mathscr{C}$ is as claimed. As for (G), note that, for all $x, y, t \in \mathbb{R}_{\geq 0}$, the ratio $\frac{f(x+y+t)-f(x)}{f(x+y+t)-f(y+t)}$ is equal to x/y if f is linear, and is equal to $e^{\phi(x-y)}$ if f is of the form $ae^{\phi x} + b$. Thus, (G) is satisfied for all x, y only if $\mathscr{C}$ is as claimed. □

5 Notes on the Literature

The priority rule for minimizing the weighted sum of completion times is due to Smith [19] and today known as *Smith's rule*. Rothkopf [17] observed that exponential cost functions also admit a simple priority rule that minimizes the total delay costs. Rothkopf and Smith [18] showed that there is no priority rule that optimizes the total delay for non-linear and non-exponential functions. Their proof uses similar ideas as ours, but requires that the cost functions are twice differentiable.

Yuan [21] proved that single-machine scheduling is weakly NP-hard for weighted tardiness delay cost functions of the form $a_j \max\{0, x - d\}$ where d is a common due date of all jobs; Höhn and Jacobs [10] showed strong NP-hardness for the case that $\mathscr{C}$ contains integer multiples of a piecewise linear and monotone function.

Although not optimal, Smith's rule gives good approximations on the minimum total delay costs for non-linear and non-exponential delay cost functions; for concave functions the approximation factor is $(\sqrt{3}+1)/2$, see Stiller and Wiese [20]; for monomials of degree k, the factor is roughly $k^{(k-1)/(k+1)}$, see Höhn and Jacobs [10]. Epstein et al. [3] show how to construct a sequence of jobs that is a 4-approximation for all delay cost functions $a_j f$, where f is a non-decreasing and non-negative function. Turning to non-universal algorithms, Megow and Verschae [13] give a polynomial time approximation scheme.

Congestion games were introduced by Rosenthal [16], who showed by a potential function argument that each game with unit demand players admits a pure Nash equilibrium. Monderer and Shapley [14] generalized this approach and showed that each exact potential game is isomorphic to an unweighted congestion game. For congestion games with weighted players, several works observed independently that a pure Nash equilibrium need not exist [4, 6, 12]. In fact, it is NP-complete to decide whether a weighted congestion game has a pure Nash equilibrium [2]. Fotakis et al. [4] and Panagopoulou and Spirakis [15] proved the existence of a pure Nash equilibrium in weighted congestion games with linear costs and exponential costs, respectively. Anshelevich et al. [1] showed that two-player games with cost functions of the form $f(x) = k/x$ with $k \in \mathbb{R}$ have a pure Nash equilibrium. These positive results rely on the construction of potential functions similar to those used in the proof of Theorem 2. Harks et al. [9] showed the limit of this approach as they proved that two-player games have a weighted potential function in general if and only if all cost functions are linear transformations of each other; and games with three or more players admit a weighted potential function in general if only if all cost functions are linear or all cost functions are exponential. Their characterization builds upon the solution of the functional equation given in Sect. 4. Harks and Klimm [7] showed that for all other sets of cost functions, a pure Nash equilibrium need not exist. Similar characterizations have been obtained for congestion games with variable or resource-dependent demands [8], congestion games with player-specific costs [5] and congestion games with multi-dimensional demands [11].

Acknowledgments Part of the results of this chapter are joint work with Tobias Harks and Rolf H. Möhring. I also wish to thank Wiebke Höhn for introducing me to the topic of priority rules for single-machine scheduling problems.

References

1. Anshelevich, E., Dasgupta, A., Kleinberg, J., Tardos, É., Wexler, T., Roughgarden, T.: The price of stability for network design with fair cost allocation. SIAM J. Comput. **38**(4), 1602–1623 (2008)
2. Dunkel, J., Schulz, A.S.: On the complexity of pure-strategy Nash equilibria in congestion and local-effect games. Math. Oper. Res. **33**, 851–868 (2008)
3. Epstein, L., Levin, A., Marchetti-Spaccamela, A., Megow, N., Mestre, J., Skutella, M., Stougie, L.: Universal sequencing on an unreliable machine. SIAM J. Comput. **41**(3), 565–586 (2012)
4. Fotakis, D., Kontogiannis, S., Spirakis, P.: Selfish unsplittable flows. Theor. Comput. Sci. **348**(2–3), 226–239 (2005)
5. Gairing, M., Klimm, M.: Congestion games with player-specific costs revisited. In: Vöcking, B. (ed.) Proceedings of the 6th International Symposium on Algorithmic Game Theory (SAGT), pp. 98–109 (2013)
6. Goemans, M., Mirrokni, V., Vetta, A.: Sink equilibria and convergence. In: Proceedings of the 46th Annual Symposium on Foundations of Computer Science (FOCS), pp. 142–154 (2005)
7. Harks, T., Klimm, M.: On the existence of pure Nash equilibria in weighted congestion games. Math. Oper. Res. **37**(3), 419–436 (2012)
8. Harks, T., Klimm, M.: Congestion games with variable demands. Math. Oper. Res. (to appear)
9. Harks, T., Klimm, M., Möhring, R.H.: Characterizing the existence of potential functions in weighted congestion games. Theory Comput. Syst. **49**(1), 46–70 (2011)
10. Höhn, W., Jacobs, T.: On the performance of Smith's rule in single-machine scheduling with nonlinear cost. ACM Trans. Algorithms **11**(4), 25 (2015)

11. Klimm, M., Schütz, A.: Congestion games with higher demand dimensions. In: Liu, T.Y., Qi, Q., Ye, Y. (eds.) Proceedings of the 10th International Conference on Web and Internet Economics (WINE), pp. 453–459 (2014)
12. Libman, L., Orda, A.: Atomic resource sharing in noncooperative networks. Telecommun. Syst. **17**(4), 385–409 (2001)
13. Megow, N., Verschae, J.: Dual techniques for scheduling on a machine with varying speed. In: Fomin, F.V., Freivalds, R., Kwiatkowska, M., Peleg, D. (eds.) Proceedings of the 40th International Colloquium on Automata, Languages and Programming (ICALP). Lecture Notes in Computer Science, vol. 7965, pp. 745–756 (2013)
14. Monderer, D., Shapley, L.S.: Potential games. Games Econ. Behav. **14**(1), 124–143 (1996)
15. Panagopoulou, P., Spirakis, P.: Algorithms for pure Nash equilibria in weighted congestion games. ACM J. Exp. Algorithmics **11**, 1–19 (2006)
16. Rosenthal, R.: A class of games possessing pure-strategy Nash equilibria. Int. J. Game Theory **2**(1), 65–67 (1973)
17. Rothkopf, M.H.: Scheduling independent tasks on parallel processors. Manag. Sci. **12**(5), 437–447 (1966)
18. Rothkopf, M.H., Smith, S.A.: There are no undiscovered priority index sequencing rules for minimizing total delay costs. Math. Oper. Res. **32**(2), 451–456 (1984)
19. Smith, W.E.: Various optimizers for single-stage production. Nav. Res. Logist. Quart. **3**(1–2), 59–66 (1956)
20. Stiller, S., Wiese, A.: Increasing speed scheduling and flow scheduling. In: Cheong, O., Chwa, K.Y., Park, K. (eds.) Proceedings of the 21st International Symposium on Algorithms and Computation (ISAAC). Lecture Notes in Computer Science, vol. 6507, pp. 279–290 (2010)
21. Yuan, J.: The NP-hardness of the single machine common due date weighted tardiness problem. Syst. Sci. Math. Sci. **5**(4), 328–333 (1992)

Convex Quadratic Programming in Scheduling

Martin Skutella

Abstract We consider the optimization problem of scheduling a given set of jobs on unrelated parallel machines with total weighted completion time objective. This is a classical scheduling problem known to be NP-hard since the 1970s. We give a new and simplified version of the currently best-known approximation algorithm, which dates back to 1998. It achieves performance ratio 3/2, and is based on an optimal solution to a convex quadratic program.

1 Problem Description

The input to our problem consists of n jobs $j = 1, 2, \ldots, n$, a number m of machines, job weights $w_j \geq 0$ and processing times $p_{ij} > 0$, for $1 \leq i \leq m$, $1 \leq j \leq n$. In a feasible schedule, every job j must be assigned to some machine i where it is being processed without interruption for time p_{ij}. As a machine may not process two or more jobs at a time, all jobs assigned to a particular machine must be sequenced and then processed one after another from time zero on. The completion time of job j in a particular schedule is denoted by C_j. The objective is to minimize the total weighted completion time $\sum_{j=1}^{n} w_j C_j$. An illustrating example is given in Fig. 1.

This parallel machine scheduling problem can be reduced to an assignment problem; for a given assignment of jobs to machines the sequencing of the assigned jobs can be done optimally on each machine i by applying Smith's ratio rule: schedule the jobs in order of non-increasing ratios w_j/p_{ij}. To keep the notation simple, we assume that $w_j/p_{ij} \neq w_k/p_{ik}$ for each machine i and each pair of jobs $j \neq k$. In particular, this implies that Smith's ratio rule yields a unique sequence of jobs for each machine. We introduce for each machine i a corresponding total order $\prec_i$ where $j \prec_i k$ holds for a pair of jobs $j \neq k$ if and only if $w_j/p_{ij} > w_k/p_{ik}$.

M. Skutella (✉)
Institut für Mathematik, Technische Universität Berlin, Straße des 17. Juni 136, 10623 Berlin, Germany
e-mail: martin.skutella@tu-berlin.de

© Springer International Publishing Switzerland 2015
A.S. Schulz et al. (eds.), *Gems of Combinatorial Optimization and Graph Algorithms*, DOI 10.1007/978-3-319-24971-1_12

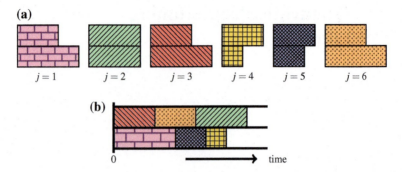

Fig. 1 (a) An instance of the unrelated parallel machine scheduling problem consisting of $m = 2$ machines and six jobs depicted in different colors and patterns. The length of the *upper/lower* bar representing each job j corresponds to j's processing time on the first/second machine, that is, p_{1j} and p_{2j}, respectively. (b) A feasible solution where, for example, the *orange dotted* job $j = 6$ is processed on the first machine and completes at time $C_6 = p_{13} + p_{16}$.

2 A Quadratic Programming Formulation

We introduce binary assignment variables

$$x_{ij} \in \{0, 1\} \quad \text{for} \quad i = 1, \ldots, m, \ j = 1, \ldots, n, \tag{1}$$

with the meaning that $x_{ij} = 1$ if and only if job j is assigned to machine i. As each job must be assigned to exactly one machine, we get the constraints

$$\sum_{i=1}^{m} x_{ij} = 1 \quad \text{for} \quad j = 1, \ldots, n. \tag{2}$$

If job j is assigned to machine i (i.e., $x_{ij} = 1$), its completion time C_j is the sum of its own processing time p_{ij} and the processing times of other jobs $k \prec_i j$ which are also scheduled on machine i. That is,

$$C_j = \sum_{i=1}^{m} x_{ij} \cdot \left(p_{ij} + \sum_{k \prec_i j} x_{ik} \cdot p_{ik} \right) \quad \text{for} \quad j = 1, \ldots, n. \tag{3}$$

Notice that the term on the right-hand side of (3) contains terms that are quadratic in x. For the subsequent discussion it is important that these quadratic terms always contain two variables x_{ij} and x_{ik} belonging to two different jobs $j \neq k$.

The integer quadratic programming formulation to minimize $\sum_{j=1}^{n} w_j C_j$ subject to (1)–(3) can be compactly rewritten by removing constraints (3) and replacing C_j in the objective function with the corresponding term on the right-hand side of (3). We refer to the resulting integer quadratic program as (IQP):

$$\text{minimize} \quad c^T x + \tfrac{1}{2} x^T D x \tag{4}$$

$$\text{subject to} \quad \sum_{i=1}^{m} x_{ij} = 1 \quad \text{for} \quad j = 1, \ldots, n \tag{2}$$

$$x \in \{0, 1\}^{mn} \tag{1}$$

Here, $x \in \{0,1\}^{mn}$ denotes the vector consisting of all variables x_{ij}, vector $c \in \mathbb{R}^{mn}$ is given by $c_{ij} := w_j p_{ij} \geq 0$, and $D = (d_{(ij)(hk)})$ is a symmetric non-negative $mn \times mn$-matrix whose entries $d_{(ij)(hk)}$, for machines i, h and jobs j, k, are given by

$$d_{(ij)(hk)} := \begin{cases} 0 & \text{if } i \neq h \text{ or } j = k, \\ w_j p_{ik} & \text{if } i = h \text{ and } k \prec_i j, \\ w_k p_{ij} & \text{if } i = h \text{ and } j \prec_i k. \end{cases} \tag{5}$$

If we relax the integrality requirements for the binary assignment variables x_{ij} and replace (1) with $x \geq 0$ (notice that $x \leq 1$ is implied by (2)), we obtain a quadratic program which we call (QP). It turns out that (IQP) and (QP) have the same optimal value. Even more, any feasible solution to (QP) can efficiently be turned into a feasible solution to (IQP) without increasing the objective function value:

Lemma 1 *Given a vector $x \geq 0$ satisfying (2), one can compute in polynomial time a feasible solution $\bar{x}$ to (IQP) with*

$$c^T \bar{x} + \tfrac{1}{2} \bar{x}^T D \bar{x} \leq c^T x + \tfrac{1}{2} x^T D x.$$

Proof For each job j choose one machine i independently at random with probability x_{ij}; set $\bar{x}_{ij} := 1$ and $\bar{x}_{i'j} := 0$ for $i' \neq i$. Notice that the random choices for different jobs $j \neq j'$ are independent so that, for each machine i,

$$\mathbb{E}[\bar{x}_{ij} \cdot \bar{x}_{ij'}] = \mathbb{E}[\bar{x}_{ij}] \cdot \mathbb{E}[\bar{x}_{ij'}] = x_{ij} \cdot x_{ij'}.$$

Taking into account that $d_{(ij)(hj)} = 0$ for $i \neq h$ and by linearity of expectation we therefore get

$$\mathbb{E}[c^T \bar{x} + \tfrac{1}{2} \bar{x}^T D \bar{x}] = c^T x + \tfrac{1}{2} x^T D x.$$

Derandomizing this randomized rounding procedure by the method of conditional expectations yields the desired result. □

It follows from Lemma 1 that it is still NP-hard to find an optimal solution to the quadratic program (QP). In the next section we convexify the objective function of (QP) in order to get a convex quadratic program, which can be solved in polynomial time.

3 A Convex Quadratic Program

In order to better understand the structure of matrix D, we assume that the entries of vectors $x = (x_{ij})$ and $c = (c_{ij})$, as well as the rows and columns of matrix D are lexicographically ordered with respect to the natural order $1, 2, \ldots, m$ of machines i and then, for each machine i, jobs j are ordered according to $\prec_i$. As a consequence of this ordering, matrix D decomposes into m diagonal blocks D_i, $i = 1, \ldots, m$, corresponding to the m machines. Furthermore, if we assume that jobs are indexed according to $\prec_i$ (i.e., $1 \prec_i 2 \prec_i \ldots \prec_i n$), the ith block D_i has the following form (cp. (5)):

$$D_i = \begin{pmatrix} 0 & w_2 p_{i1} & w_3 p_{i1} & \cdots & w_n p_{i1} \\ w_2 p_{i1} & 0 & w_3 p_{i2} & \cdots & w_n p_{i2} \\ w_3 p_{i1} & w_3 p_{i2} & 0 & & w_n p_{i3} \\ \vdots & \vdots & & \ddots & \vdots \\ w_n p_{i1} & w_n p_{i2} & w_n p_{i3} & \cdots & 0 \end{pmatrix}. \qquad (6)$$

Notice that all diagonal entries of matrix D_i are zero. Moreover, assuming there is at least one job j with strictly positive weight w_j, matrix D_i has off-diagonal entries that are non-zero. Therefore, the matrix is not positive semidefinite and the quadratic objective function (4) is not convex. Let $\mathrm{diag}(c)$ denote the diagonal $mn \times mn$-matrix whose diagonal entries coincide with the entries of vector c. The following lemma indicates how (4) can be convexified.

Lemma 2 *The matrix $D + \mathrm{diag}(c)$ is positive semidefinite.*

Proof Using the same notation as in (6), the ith block of $D + \mathrm{diag}(c)$ has the form:

$$A_i := \begin{pmatrix} w_1 p_{i1} & w_2 p_{i1} & w_3 p_{i1} & \cdots & w_n p_{i1} \\ w_2 p_{i1} & w_2 p_{i2} & w_3 p_{i2} & \cdots & w_n p_{i2} \\ w_3 p_{i1} & w_3 p_{i2} & w_3 p_{i3} & \cdots & w_n p_{i3} \\ \vdots & \vdots & \vdots & \ddots & \vdots \\ w_n p_{i1} & w_n p_{i2} & w_n p_{i3} & \cdots & w_n p_{in} \end{pmatrix}. \qquad (7)$$

We prove that matrix A_i is positive semidefinite by showing that the determinants of all its principal sub-matrices are non-negative. Note that each principal sub-matrix corresponds to a subset of jobs and is of the same form as A_i for the smaller instance induced by this subset of jobs. Therefore it suffices to show that the determinant of A_i is non-negative for all instances. The proof uses induction on the number of jobs n. The case $n = 1$ is trivial. For $n > 1$, by subtracting p_{i1}/p_{i2} times the second column of A_i from its first column, we get

$$\det A_i = \det \begin{pmatrix} p_{i1}^2(\frac{w_1}{p_{i1}} - \frac{w_2}{p_{i2}}) & w_2 p_{i1} & w_3 p_{i1} & \cdots & w_n p_{i1} \\ 0 & w_2 p_{i2} & w_3 p_{i2} & \cdots & w_n p_{i2} \\ 0 & w_3 p_{i2} & w_3 p_{i3} & \cdots & w_n p_{i3} \\ \vdots & \vdots & \vdots & \ddots & \vdots \\ 0 & w_n p_{i2} & w_n p_{i3} & \cdots & w_n p_{in} \end{pmatrix}$$

$$= p_{i1}^2 \left(\frac{w_1}{p_{i1}} - \frac{w_2}{p_{i2}} \right) \cdot \det \begin{pmatrix} w_2 p_{i2} & w_3 p_{i2} & \cdots & w_n p_{i2} \\ w_3 p_{i2} & w_3 p_{i3} & \cdots & w_n p_{i3} \\ \vdots & \vdots & \ddots & \vdots \\ w_n p_{i2} & w_n p_{i3} & \cdots & w_n p_{in} \end{pmatrix}.$$

As $p_{i1}^2 > 0$ and $\frac{w_1}{p_{i1}} \geq \frac{w_2}{p_{i2}}$, this concludes the proof by induction. □

Lemma 2 leads to the following convex quadratic program (CQP):

$$\text{minimize} \quad c^T x + \tfrac{1}{2} x^T (D + \text{diag}(c)) x$$
$$\text{subject to} \quad \sum_{i=1}^{m} x_{ij} = 1 \quad \text{for } j = 1, \ldots, n$$
$$x \geq 0$$

Since (CQP) can be solved in polynomial time, we obtain an approximation algorithm for our unrelated parallel machine scheduling problem as follows:

Theorem 3 *Computing an optimal (fractional) solution x to (CQP) and turning it into a feasible solution $\bar{x}$ to (IQP) according to Lemma 1 is a 3/2-approximation algorithm.*

Proof In order to prove the performance ratio, notice that, according to Lemma 1, the value of the solution $\bar{x}$ is bounded by

$$c^T \bar{x} + \tfrac{1}{2} \bar{x}^T D \bar{x} \leq c^T x + \tfrac{1}{2} x^T D x \leq c^T x + \tfrac{1}{2} x^T (D + \text{diag}(c)) x. \quad (8)$$

The second inequality holds because $x^T (D + \text{diag}(c)) x = x^T D x + x^T \text{diag}(c) x$ and $x^T \text{diag}(c) x \geq 0$ by non-negativity of x and c. Recall that x minimizes the term on the right-hand side of (8). In particular, for an optimal solution x^* to (IQP) we obtain

$$c^T x + \tfrac{1}{2} x^T (D + \text{diag}(c)) x \leq c^T x^* + \tfrac{1}{2} x^{*T} (D + \text{diag}(c)) x^*. \quad (9)$$

Finally, as x^* has binary entries, $x^{*T} \text{diag}(c) x^* = c^T x^*$ and we get

$$c^T x^* + \tfrac{1}{2} x^{*T} (D + \text{diag}(c)) x^* = \tfrac{3}{2} c^T x^* + \tfrac{1}{2} x^{*T} D x^* \leq \tfrac{3}{2} \left(c^T x^* + \tfrac{1}{2} x^{*T} D x^* \right), \quad (10)$$

where the last inequality follows by non-negativity of x^* and D. □

Notice that (CQP) is not a relaxation of the minimization problem (IQP) because, for a feasible solution to (IQP), the objective function of (CQP) can be strictly larger than the objective function of (IQP). On the other hand, it follows from inequalities (9) and (10) in the proof of Theorem 3 that scaling the convex quadratic objective function of (CQP) by a factor 2/3 yields a lower bound and therefore a convex quadratic programming relaxation of the scheduling problem under consideration. In the following, we refer to this relaxation as (CQPR).

4 The Special Case of Identical Parallel Machines

We can simplify the approximation result for the case of identical parallel machines, that is, when $p_{1j} = \ldots = p_{mj} =: p_j$ for each job j.

Lemma 4 *For the case of identical parallel machines an optimal solution to (CQP) is given by $x_{ij} := \frac{1}{m}$ for all i, j.*

Proof Let $x' \neq x$ be a feasible solution to (CQP). Since (CQP) is symmetric with respect to the m identical machines, we get $m - 1$ additional solutions of the same value as x' by cyclically permuting the machines $m - 1$ times. The convex combination with coefficients $\frac{1}{m}$ of x' and these new solutions is precisely x. Since the objective function of (CQP) is convex, the value of x is less than or equal to the value of x'. $\square$

Using the randomized rounding procedure described in the proof of Lemma 1 based on this optimal solution x to (CQP) thus yields the following result, which is a special case of Theorem 3.

Corollary 5 *Assigning jobs independently at random to machines is a randomized 3/2-approximation algorithm.*

It is not difficult to see that, even for the special case of identical parallel machines, the optimal solution value of the convex quadratic programming relaxation (CQPR) discussed at the end of the last section can be smaller than the optimal value of (IQP) by a factor $(2 + \frac{1}{m})/3$. It suffices to consider an instance consisting of a single job and m identical parallel machines.

One can also show that the performance guarantee given in Theorem 3 and Corollary 5 is asymptotically tight. Consider an instance consisting of m identical parallel machines and m jobs of weight $1/m$ and unit processing time. An optimal solution has value 1 while the expected value of the solution described in Corollary 5 is $\frac{3}{2} - \frac{1}{2m}$. A slightly more detailed analysis shows that the performance guarantee in Corollary 5 can indeed be improved to $\frac{3}{2} - \frac{1}{2m}$.

5 Notes on the Literature

Smith [16] observed that minimizing the total weighted completion time on a single machine is easy. (A result now known as Smith's ratio rule.) Already for the case of two identical parallel machines the problem is NP-hard in the ordinary sense [1]. If the number m of identical parallel machines is part of the input, the problem is strongly NP-hard; see, for example, the classical book of Garey and Johnson [2, Problem SS13].

The presented 3/2-approximation algorithm for the unrelated parallel machines scheduling problem is a simplified version of an approximation result that was first presented by the author in [11, 12] and later, independently, by Sethuraman and Squillante in [9]. In [9, 11, 12] the following slightly stronger convex quadratic programming relaxation is used:

$$\begin{aligned}
\text{minimize} \quad & Z \\
\text{subject to} \quad & \sum_{i=1}^{m} x_{ij} = 1 \quad \text{for } j = 1, \ldots, n \\
& Z \geq c^T x \quad &(11) \\
& Z \geq \tfrac{1}{2} c^T x + \tfrac{1}{2} x^T (D + \text{diag}(c)) x \quad &(12) \\
& x \geq 0
\end{aligned}$$

As an optimal solution to this program might be irrational (see [14]), it can only be solved up to an additive error of $\varepsilon > 0$ in polynomial time. This issue somewhat complicates the proof of the performance guarantee 3/2. In contrast, the convex quadratic programs (CQP) and (CQPR) introduced in Sect. 3 above can be solved to optimality in polynomial time since only their objective functions are quadratic and all constraints are linear. Notice that the objective function of the convex quadratic programming relaxation (CQPR) is a convex combination of the right-hand sides of (11) and (12) with coefficients 1/3 and 2/3, respectively.

Hoogeveen, Schuurman, and Woeginger [3] showed that, unless P = NP, there is no polynomial-time approximation scheme for the scheduling problem under consideration. It is an interesting and long-standing open problem to close the gap between this lower bound and the approximation results discussed above. According to Schuurman and Woeginger [7], this is one of the ten most notable open questions in the area of approximation algorithms for machine scheduling problems.

Convex quadratic programs have proved useful in approximating other and more general machine scheduling problems, including, e.g., preemptions and release dates; see [10, 13, 14] for examples.

The result in Corollary 5 on the special case of identical parallel machines was first obtained in [5], based on a closely related time-indexed linear programming relaxation; see also [6, 11]. Quite interestingly, the derandomization of this randomized algorithm precisely coincides with the WSPT-rule: list the jobs according to non-increasing ratios w_j/p_j and schedule the next job whenever a machine becomes available. Kawaguchi and Kyan [4] (see also Schwiegelshohn [8]) showed that the WSPT-rule even achieves performance ratio $(1+\sqrt{2})/2$. The first polynomial-time approximation scheme for this identical parallel machine scheduling problem was presented in [15].

References

1. Bruno, J.L., Coffman Jr. E.G., Sethi, R.: Scheduling independent tasks to reduce mean finishing time. Commun. Assoc. Comput. Mach. **17**, 382–387 (1974)
2. Garey, M.R., Johnson, D.S.: Computers and Intractability: A Guide to the Theory of NP-Completeness. Freeman, San Francisco (1979)
3. Hoogeveen, H., Schuurman, P., Woeginger, G.J.: Non-approximability results for scheduling problems with minsum criteria. INFORMS J. Comput. **13**, 157–168 (2001)
4. Kawaguchi, T., Kyan, S.: Worst case bound of an LRF schedule for the mean weighted flow-time problem. SIAM J. Comput. **15**, 1119–1129 (1986)
5. Schulz, A.S., Skutella, M.: Random-based scheduling: New approximations and LP lower bounds. In: Rolim, J. (ed.) Randomization and Approximation Techniques in Computer Science. Lecture Notes in Computer Science, vol. 1269, pp. 119–133. Springer (1997)
6. Schulz, A.S., Skutella, M.: Scheduling unrelated machines by randomized rounding. SIAM J. Discret. Math. **15**, 450–469 (2002)
7. Schuurman, P., Woeginger, G.J.: Polynomial time approximation algorithms for machine scheduling: Ten open problems. J. Sched. **2**, 203–213 (1999)
8. Schwiegelshohn, U.: An alternative proof of the Kawaguchi-Kyan bound for the largest-ratio-first rule. Oper. Res. Lett. **39**, 255–259 (2011)
9. Sethuraman, J., Squillante, M.S.: Optimal scheduling of multiclass parallel machines. In: Proceedings of the 10th Annual ACM-SIAM Symposium on Discrete Algorithms, pp. 963–964, Baltimore (1999)
10. Sitters, R.A.: Approximability of average completion time scheduling on unrelated machines. In: Halperin, D., Mehlhorn, K. (eds.) Algorithms–ESA'08. Lecture Notes in Computer Science, vol. 5193, pp. 768–779. Springer (2008)
11. Skutella, M.: Approximation and Randomization in Scheduling. Ph.D. thesis, Technische Universität Berlin, Germany (1998)
12. Skutella, M.: Semidefinite relaxations for parallel machine scheduling. In: Proceedings of the 39th Annual IEEE Symposium on Foundations of Computer Science, pp. 472–481. Palo Alto (1998)
13. Skutella, M.: Convex quadratic programming relaxations for network scheduling problems. In: Nešetřil, J. (ed.) Algorithms–ESA'99. Lecture Notes in Computer Science, vol. 1643, pp. 127–138. Springer (1999)
14. Skutella, M.: Convex quadratic and semidefinite programming relaxations in scheduling. J. ACM **48**, 206–242 (2001)
15. Skutella, M., Woeginger, G.J.: A PTAS for minimizing the total weighted completion time on identical parallel machines. Math. Oper. Res. **25**, 63–75 (2000)
16. Smith, W.E.: Various optimizers for single-stage production. Nav. Res. Log. Q. **3**, 59–66 (1956)

Robustness and Approximation for Universal Sequencing

Nicole Megow

Abstract We consider the problem of finding a permutation of jobs that minimizes $\sum_j w_j f(C_j)$ on a single machine for some non-negative, non-decreasing global cost function f. We are interested in universal solutions that perform well for all functions f simultaneously. We construct universal sequences that are within a factor of 4 of the optimal cost for any f. Furthermore, we analyze the tradeoff between the robustness for *all* cost functions and the approximation of the well understood case of *linear* cost functions.

1 Problem Description

We are given a set of jobs $J = \{1, 2, \ldots, n\}$ with processing times $p_j > 0$ and weights $w_j > 0$, $j \in J$. Using a standard scaling argument, we may assume $w_j \geq 1$. A feasible schedule on a single machine is given by a permutation of jobs. Given a permutation π, let $C_j^\pi := \sum_{k:\pi(k) \leq \pi(j)} p_k$ denote the completion time of job j when scheduling all jobs without preemption and without inserting idle time in order of π. The task is to find a permutation that minimizes the total cost $\sum w_j f(C_j^\pi)$, where $f : \mathbb{R}_{\geq 0} \to \mathbb{R}_{\geq 0}$ is a non-decreasing (and thus almost everywhere differentiable) cost function with $f(0) = 0$. In standard scheduling notation this problem is denoted as $1 || \sum w_j f(C_j)$. We denote the total cost of an optimal permutation for cost function f by $\text{OPT}(f)$. We assume that f is given implicitly by an oracle which returns the cost for a given completion time. We are interested in *universal sequences* that perform well for all cost functions f, simultaneously.

N. Megow (✉)
Fakultät für Mathematik, Technische Universität München, Boltzmannstr. 3, 85748 Garching, Germany
e-mail: nmegow@ma.tum.de

© Springer International Publishing Switzerland 2015
A.S. Schulz et al. (eds.), *Gems of Combinatorial Optimization and Graph Algorithms*, DOI 10.1007/978-3-319-24971-1_13

Definition 1 For $\alpha \geq 1$, a sequence π is *α-robust* if, for all cost functions f, the total cost of π is at most α times the cost of the optimal schedule for function f. That is, for any f we have

$$\sum_{j \in J} w_j f(C_j^\pi) \leq \alpha \, \text{OPT}(f).$$

2 The Min-Sum Objective and a Lower Bound

Given a sequence π, let $\chi_j^\pi(t)$ be the indicator function, which is 1 if and only if job j is unfinished at time t in the schedule according to π. The cost for completing job j is

$$f(C_j^\pi) = \int_0^{C_j^\pi} f'(t)dt = \int_0^\infty \chi_j^\pi(t) f'(t)dt.$$

For some point in time $t \geq 0$, let $W^\pi(t)$ denote the total weight of jobs that are not yet completed by time t according to π, that is, $W^\pi(t) := \sum_{j \in J} \chi_j^\pi(t) w_j$. Then,

$$\sum_{j \in J} w_j f(C_j^\pi) = \int_0^\infty W^\pi(t) f'(t)dt. \tag{1}$$

For $t \geq 0$, let $W^*(t) := \min_\pi W^\pi(t)$ be the minimum total weight of unfinished jobs in the system at time t in any schedule. With this definition we can directly give a lower bound on the scheduling cost of any permutation, and thus, also the optimal solution.

Observation 2 *The objective value $\sum_{j \in J} w_j f(C_j)$ of any schedule is bounded from below by*

$$\int_0^\infty W^*(t) f'(t)dt.$$

The following lemma states a relation between robustness and the remaining-weight functions W and W^* independently of the global cost function f.

Lemma 3 *A job sequence π is α-robust if $W^\pi(t) \leq \alpha W^*(t)$ for all times $t \geq 0$.*

Proof If $W^\pi(t) \leq \alpha W^*(t)$, for any t, then Eq. (1) and Observation 2 directly imply

$$\sum_{j \in J} w_j f(C_j^\pi) = \int_0^\infty W^\pi(t) f'(t)dt \leq \alpha \int_0^\infty W^*(t) f'(t)dt \leq \alpha \, \text{OPT}(f).$$

$\square$

This key observation gives a sufficient condition for α-robust sequencing, which is independent of the actual cost function f. For approximating the min-sum cost for *any* f, we only need to approximate the total weight of uncompleted jobs at any point in time.

3 A Universal Sequencing Algorithm

We present an algorithm that determines a universal sequence that is 4-robust. By Lemma 3, it is sufficient to give point-wise a relative guarantee on the remaining-weight value. Thus, our algorithm aims at minimizing this value at any point in time. Intuitively, the end of the sequence is more critical because the remaining-weight function values are small. Therefore, our algorithm computes the sequence iteratively backwards. In each iteration it finds a set of jobs that has maximum total processing time subject to a bound on the total weight of this set. This weight bound is doubled in each iteration. For the analysis it is crucial that we solve the subset selection problem in each iteration independently. That means, we select from the set of *all* jobs, including those that have been selected in an earlier iteration.

We use the notation $p(A) := \sum_{j \in A} p_j$ and $w(A) := \sum_{j \in A} w_j$, for any $A \subseteq J$.

Algorithm 1 DOUBLING(J)

1. $\pi \leftarrow$ empty sequence
2. **for** $i \in \{0, 1, \ldots, \lceil \log_2 w(J) \rceil\}$ **do**
3. $J_i \leftarrow$ subset of all jobs J maximizing $p(J_i)$ subject to $w(J_i) \leq 2^i$.
4.   prepend to π the jobs $J_i \setminus \bigcup_{k=0}^{i-1} J_k$ in arbitrary order
5. **return** π

Theorem 4 *For every scheduling instance,* DOUBLING *produces a universal 4-robust schedule.*

Proof By Lemma 3 it is sufficient to show that $W^\pi(t) \leq 4W^*(t)$ for all $t \geq 0$. Let $t \geq 0$, and let i be minimal such that $p(J_i) \geq p(J) - t$. By construction of π, only jobs j in $\bigcup_{k=0}^{i} J_k$ can have a completion time $C_j^\pi > t$. Thus,

$$W^\pi(t) \leq \sum_{k=0}^{i} w(J_k) \leq \sum_{k=0}^{i} 2^k = 2^{i+1} - 1. \tag{2}$$

In case $i = 0$, the claim is trivially true since $w_j \geq 1$ for any $j \in J$, and thus, $W^*(t) = W^\pi(t)$. Suppose $i \geq 1$, then by our choice of i, it holds that $p(J_{i-1}) < p(J) - t$. Therefore, in any sequence π', the total weight of jobs completing after time t is larger than 2^{i-1}, because otherwise we get a contradiction to the maximality of $p(J_{i-1})$. That is, $W^*(t) > 2^{i-1}$. Together with (2) this concludes the proof. □

We remark that finding the subsets of jobs J_i in Line 3 of Algorithm DOUBLING is a KNAPSACK problem and, thus, NP-hard. Using straightforward dynamic programming, our algorithm runs in pseudo-polynomial time and achieves a performance guarantee of 4 as shown above. However, known FPTASes for KNAPSACK can be adopted such that DOUBLING runs in polynomial time, while loosing only an arbitrarily small constant in the robustness factor.

It is possible to obtain a better robustness factor if we select the sequence at random and slightly relax the universality requirement.

Definition 5 A probability distribution over sequences π is a *randomized α-robust* solution if for any cost function f the *expected* total cost of a sequence π chosen according to the distribution is at most α times the total cost $\text{OPT}(f)$ of an optimal schedule for this particular function f. That is, for any f we have

$$\mathbb{E}\left[\sum_{j \in J} w_j f(C_j^\pi)\right] \leq \alpha \, \text{OPT}(f).$$

The deterministic Algorithm DOUBLING computes in each iteration an optimal subset of jobs meeting a given weight bound. This weight bound is 2^i in iteration i, that is, it is *doubled* in each iteration. In a natural randomized variant of this algorithm, we increase the weight bound by a randomly chosen factor instead of 2. We can choose the probability distribution in such a way that the the algorithm achieves a robustness factor of $e \approx 2.718$.

Theorem 6 *For every scheduling instance, a randomized variant of* DOUBLING *produces a randomized e-robust solution.*

4 Worst-Case Versus Best-Case

Robust solutions are very useful in uncertainty environments in which the cost function is not or only partially known. However, the robustness for *all* cost functions has its price when facing a well-understood *best-case* cost function. Consider the global cost function to be some linear function g. This case is of particular interest, because it may be the standard setting, and most importantly, it can be handled optimally by simply ordering the jobs in non-increasing order of weight over processing time ratio, called *Smith's Rule*. However, the schedule produced by DOUBLING is usually suboptimal. It may be hard to sell to a practitioner an α-robust solution (even if α is bounded) that deviates a lot from the optimal solution in the standard setting.

This motivates the study of the tradeoff between worst-case robustness and best-case approximability. More specifically, we would like to determine pairs of (α, β) for which it possible to find a permutation π that is α-robust (for all cost functions) and β-approximate for linear cost functions g. For the ease of exposition, we restrict here to the special case in which the standard cost function g is the identity function $g(t) = t$, for all $t \geq 0$.

Definition 7 A job sequence π is β-*approximate* if the total cost of π when the global cost function is the identity function is at most β times the cost of the optimal sequence σ following Smith's Rule, that is, we have

$$\sum_{j \in J} w_j C_j^\pi \leq \beta \sum_{j \in J} w_j C_j^\sigma.$$

We propose and analyze the following generalization of the Algorithm DOUBLING. The algorithm takes an additional parameter $\rho > 1$ which determines the factor by which the weight bound is increased ("doubled") in each iteration. Furthermore, we order the jobs within a group more carefully using Smith's Rule.

Algorithm 2 GENERALIZED-DOUBLING(J, ρ)
---
1. $\pi \leftarrow$ empty sequence
2. **for** $i \in \{0, 1, \ldots, \lceil \log_\rho w(J) \rceil\}$ **do**
3. $J_i \leftarrow$ subset of all jobs J maximizing $p(J_i)$ subject to $w(J_i) \leq \rho^i$.
4.   sort the jobs J_i according to Smith's Rule (non-increasing order of w_j/p_j)
5.   prepend to π the jobs $J_i \setminus \bigcup_{k=0}^{i-1} J_k$ in this order
6. **return** π

Notice that Theorem 4 implies that GENERALIZED-DOUBLING with $\rho = 2$ produces a 4-robust job sequence. Since the linear cost function g is a valid cost function, we know that the schedule is 4-approximate. For this to be true, we do not even need to sort the jobs according to Smith's Rule within each group J_i.

The following theorem improves and generalizes this observation.

Theorem 8 *Let $\rho > 1$. The algorithm* GENERALIZED-DOUBLING(J, ρ) *outputs a sequence that is $\frac{\rho^2}{\rho-1}$-robust and $\left(1 + \frac{\rho}{\rho-1}\right)$-approximate.*

We prove the two bounds in the theorem separately. Consider some instance J, a linear, non-decreasing cost function f and some $\rho > 1$. Let π be the permutation computed by the algorithm and $T = p(J)$.

Lemma 9 *Solution π is $\frac{\rho^2}{\rho-1}$-robust.*

Proof To show the robustness factor, we adapt the proof of Theorem 4. For $t \geq 0$, let i be such that $p(J_i) \geq T - t > p(J_{i-1})$. It follows that $W^*(t) > \rho^{i-1}$, otherwise $p(J_{i-1}) \geq T - t$. On the other hand, in the schedule π only jobs j in $\bigcup_{k=0}^{i} J_k$ can remain unfinished at time t, that is, $C_j^\pi > t$. Therefore, we have

$$W^\pi(t) \leq \sum_{k=0}^{i} w(J_k) \leq \sum_{k=0}^{i} \rho^k = \frac{\rho^{i+1} - 1}{\rho - 1} < \frac{\rho^2}{\rho - 1} W^*(t).$$

The desired robustness factor follows directly from Lemma 3. □

Let σ be an optimal sequence for the standard cost function $g(t) = t$, $t \geq 0$, which is obtained by Smith's Rule. To prove the approximation bound, we compare the total weighted completion time of π, that is, $\sum_{j \in J} w_j C_j^\pi$, with the optimal cost $\sum_{j \in J} w_j C_j^\sigma$. Using the reformulation of the objective function, Eq. (1), we need to show

$$\int_0^T W^\pi(t) dt \leq \left(1 + \frac{\rho}{\rho - 1}\right) \int_0^T W^\sigma(t) dt . \tag{3}$$

Unlike for the robustness factor (Lemma 9), we will not be able to give a point-wise bound that holds for all $t \geq 0$. There are examples where $W^\pi(t) > \left(1 + \frac{\rho}{\rho-1}\right) W^\sigma(t)$ for some $t \geq 0$. To see that, consider the following simple instance consisting of two jobs with the parameters $w_1 = \rho$, $p_1 = \rho$ and $w_2 = 1 + 1/\rho$, $p_2 = 1 + 2/\rho$. Then, for any $\rho > 2$, our algorithm produces the sequence $\pi = (2, 1)$ because in Iteration 1 each of the jobs satisfies the weight bound of ρ, and Job 1 has the larger processing time and is selected. Job 2 is prepended in Iteration 2. Thus, at time $t = \rho + 1$, our solution has a remaining weight of $W^\pi(t) = \rho$, whereas Smith's Rule yields the reverse sequence $\sigma = (1, 2)$ with $W^\sigma(t) = 1 + 1/\rho$. Simple calculations show that $W^\pi(t)/W^\sigma(t) = \rho/(1 + \frac{1}{\rho}) > 1 + \frac{\rho}{\rho-1}$, for $\rho \geq 3.22$, which shows that the desired approximation bound does not hold at t.

Clearly, this example does not rule out the approximation factor. It only shows that we need a more refined analysis as we present below.

Lemma 10 *Solution π is $\left(1 + \frac{\rho}{\rho-1}\right)$-approximate.*

Proof To show that Inequality (3) holds true, we split the total cost $\int_0^T W^\pi(t) dt$ of π into two components and bound each of them separately. The components are determined by the grouping of jobs in the algorithm: Let $B_i = J_i \setminus \bigcup_{k=0}^{i-1} J_k$ denote the set of jobs (we also say group) that is scheduled in iteration $i \in \{0, 1, \ldots, \lceil \log_\rho w(J) \rceil\}$. For any $t \geq 0$ at which some job of group B_i is being processed, let $W_F^\pi(t)$ denote the total weight of all jobs completing later than the jobs of the group B_i that is currently being processed, that is, $W_F^\pi(t) = w(\bigcup_{k=0}^{i-1} B_k)$. Furthermore, we define $W_C^\pi(t)$ to be the total weight of jobs in group B_i with completion time greater than t. Notice that we have the relation $W^\pi(t) = W_F^\pi(t) + W_C^\pi(t)$. Hence, the total cost of π is

$$\int_0^T W^\pi(t) dt = \int_0^T W_F^\pi(t) dt + \int_0^T W_C^\pi(t) dt . \tag{4}$$

This relation is also visualized in Fig. 1. In the left picture, the red area is $\int_0^T W_F^\pi(t) dt$ and the blue area corresponds to $\int_0^T W_C^\pi(t) dt$.

In the following we will bound the two integrals in (4) separately. We claim that

$$\int_0^T W_F^\pi(t) dt \leq \frac{\rho}{\rho - 1} \int_0^T W^\sigma(t) dt \tag{5}$$

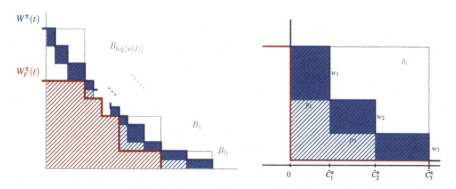

Fig. 1 Visualization of the contribution of jobs and groups to the weight function $W^\pi(t)$

and that

$$\int_0^T W_{\bar{C}}^\pi(t)\,dt \leq \int_0^T W^\sigma(t)\,dt. \tag{6}$$

The approximation factor of $1 + \frac{\rho}{\rho-1}$ follows directly from Inequalities (4)–(6).

To prove Inequality (5) we use a similar argument as in the proof of the robustness bound in Lemma 9. For any given t, let i be such that $p(J_i) \geq T - t > p(J_{i-1})$. It follows that $W^\sigma(t) \geq W^*(t) > \rho^{i-1}$. On the other hand, schedule π processes at time t some job of group B_ℓ with $\ell \leq i$. It follows that

$$W_F^\pi(t) \leq \sum_{k=0}^{i-1} w(J_k) \leq \sum_{k=0}^{i-1} \rho^k < \frac{\rho^i}{\rho - 1} < \frac{\rho}{\rho - 1} W^\sigma(t),$$

which implies (5).

To prove (6) we consider the contribution of each individual group to the total cost $\int_0^T W_{\bar{C}}^\pi(t)\,dt$. Using the reinterpretation of total scheduling cost in Eq. (1), the contribution of a single group corresponds to the total weighted completion time, when running this group of jobs on a machine of its own. More formally, for a job j of a group B_i, $i \in \{0, 1, \ldots, \lceil \log_\rho w(J) \rceil\}$, let $\bar{C}_j^\pi$ denote the completion time when scheduling only B_i according to Smith's Rule. See again Fig. 1 for a visualization. Then the total contribution of the group is exactly $\sum_{j \in B_i} w_j \bar{C}_j^\pi$. Within each group, jobs are scheduled optimally according to Smith's Rule and, thus, they follow the same relative order as σ. Hence, $\bar{C}_j^\pi \leq C_j^\sigma$. (Essentially, we say here that the total cost of an optimal parallel-machine schedule is a lower bound on an optimal single machine schedule.) It follows that

$$\int_0^T W_{\bar{C}}^\pi(t)\,dt = \sum_{i=0}^{\lceil \log_\rho w(J) \rceil} \sum_{j \in B_i} w_j \bar{C}_j^\pi \leq \sum_{i=0}^{\lceil \log_\rho w(J) \rceil} \sum_{j \in B_i} w_j C_j^\sigma = \sum_j w_j C_j^\sigma = \int_0^T W^\sigma(t)\,dt.$$

This concludes the proof. □

Lemmas 9 and 10 imply that algorithm GENERALIZED-DOUBLING (J, ρ) constructs a solution with total cost at most a factor $\rho^2/(\rho - 1)$ larger than an optimal solution for any cost function f and at most a factor $1 + \rho/(\rho - 1)$ larger than the optimal Smith Rule solution for the best-case cost function g being the identity. Thus, Lemmas 9 and 10 imply Theorem 8.

5 References to the Literature

The robustness results on universal sequences in Sect. 3 are joint work of the author with Epstein, Levin, Marchetti-Spaccamela, Mestre, Skutella and Stougie [2]. The quantification of the tradeoff between robustness and approximation of an ideal scenario is work in progress of the author and Mestre [6].

Both presented algorithms implement the idea of *guessing and doubling* an unknown parameter, which is in this case the value of the remaining-weight function at certain time points. Such type of strategies have been applied successfully for various problems; Chrobak and Kenyon-Mathieu [1] give an excellent survey with a collection of such examples. Closest to our approach is an algorithm of Hall, Schulz, Shmoys, and Wein [3] for online scheduling to minimize $\sum w_j C_j$, where the doubling happens in the time horizon.

The robustness factors for the deterministic and randomized variants of Algorithm DOUBLING shown by Epstein et al. [2] are best possible in the sense that there are instances for which these factors are tight. To address the fact that many instances clearly admit a much better robustness factor than the overall worst case, the author and Mestre [5] propose to find solutions with the best robustness factor on an instance-by-instance basis. They give a polynomial-time approximation scheme (PTAS) that approximates, for any instance, the best universal schedule up to any desired level of accuracy in polynomial time.

The problem setting in which the global cost function f is known in advance, has also been studied extensively. A PTAS has been presented recently by the author and Verschae in [7]. The performance of the well-known Smith Rule [9] has been bounded for concave cost functions in [10]. A tight analysis of the exact worst-case guarantee for Smith's rule for any convex or concave function is given in [4]. While for several classes of functions f the sequencing problem is known to be (weakly) NP-hard [4, 7, 11], only the cases with f being linear or exponential are known to be in P [8, 9]. The complexity remains open for monomials and concave cost functions.

The sequencing problem to minimize $\sum_j w_j f(C_j)$ for some global cost function f has an important interpretation in the context of scheduling on a machine of *varying speed*. In such a machine setting, we assume that the speed of the machine can change during the execution of the schedule. The problem $1||\sum_j w_j f(C_j)$ is equivalent to the problem of scheduling to minimize $\sum_j w_j C_j$ on a machine of varying speed where the speed function is chosen such that $f(C)$ equals the time that the varying-speed machine needs to process a work volume of C [4]. The special cases of concave or convex global cost functions correspond to speed functions in which the speed is non-decreasing or non-increasing, respectively. Linear cost functions, considered as the best case in Sect. 4, correspond to scheduling on a machine that is processing continuously at constant speed. An *unreliable machine* is a machine whose speed varies unpredictably during the execution of the schedule, that is, the machine capacity function is not known in advance. The universal sequencing problem to minimize $\sum_j w_j f(C_j)$ is equivalent to the problem of scheduling on an unreliable machine to minimize $\sum_j w_j C_j$. In particular, a job sequence that is α-robust for all cost functions f achieves for any machine speed function a total weighted completion time at most a factor α larger than the optimal solution for this speed function.

References

1. Chrobak, M., Kenyon-Mathieu, C.: SIGACT news online algorithms column 10: competitiveness via doubling. SIGACT News **37**(4), 115–126 (2006)
2. Epstein, L., Levin, A., Marchetti-Spaccamela, A., Megow, N., Mestre, J., Skutella, M., Stougie, L.: Universal sequencing on a single unreliable machine. SIAM J. Comput. **41**(3), 565–586 (2012)
3. Hall, L.A., Schulz, A.S., Shmoys, D.B., Wein, J.: Scheduling to minimize average completion time: Off-line and on-line approximation algorithms. Math. Oper. Res. **22**, 513–544 (1997)
4. Höhn, W., Jacobs, T.: On the performance of Smith's rule in single-machine scheduling with nonlinear cost. ACM Trans. Algorithms **11**(4) 25 (2015)
5. Megow, N., Mestre, J.: Instance-sensitive robustness guarantees for sequencing with unknown packing and covering constraints. In: Proceedings of the 4th Conference on Innovations in Theoretical Computer Science (ITCS 2013), pp. 495–504 (2013)
6. Megow, N., Mestre, J.: On the tradeoff between robustness and approximation. Manuscript (2015)
7. Megow, N., Verschae, J.: Dual techniques for scheduling on a machine with varying speed. In: Proceedings of the 40th International Colloquium on Automata, Languages and Programming (ICALP 2013), Lecture Notes in Computer Science, vol. 7965, pp. 745–756. Springer (2013)
8. Rothkopf, M.H.: Scheduling independent tasks on parallel processors. Manag. Sci. **12**, 437–447 (1966)
9. Smith, W.E.: Various optimizers for single-stage production. Nav. Res. Logist. Q. **3**, 59–66 (1956)
10. Stiller, S., Wiese, A.: Increasing speed scheduling and flow scheduling. In: Proceedings of the 21st Symposium on Algorithms and Computation (ISAAC 2010), Lecture Notes in Computer Science, vol. 6507, pp. 279–290. Springer (2010)
11. Yuan, J.: The NP-hardness of the single machine common due date weighted tardiness problem. Syst. Sci. Math. Sci. **5**(4), 328–333 (1992)

A Short Note on Long Waiting Lists

Sebastian Stiller

Abstract A waiting list is a commitment to pack a knapsack by a linear order—a commitment that has to be made before the capacity of the knapsack is known. The goal of this paper is to create a waiting list such that for every capacity the resulting knapsack solution is provably good. We show that there are waiting lists that yield at least 2-approximate knapsack solutions for every capacity. Their construction and the proof of their quality are arguably simple. Simple, but effective, as, in general, a factor better than 2 is impossible.

1 Question

What Is a Waiting List?

> Did you get tickets for the Stones?
>
> No, but we are number 68 on the waiting list.
>
> What satisfaction can you get from that?

Waiting lists are constructed when the available amount of a desired good or service is not yet known. Typical examples are stand-by capacities on airplanes, containers, hotel rooms—or tickets for a show. Once the capacity is known the waiting list determines the way the service is distributed. A group that holds number 68 on the waiting list is offered the remaining available capacity after all groups with a smaller number have been offered to take their share, and before any group with a higher number gets a chance.

Why Number 2 on a List Can Be Lucky

In 2016 nobody wants to see the Stones without the circle of friends who went together back in the 70s. Accordingly, the position r on the waiting list applies to the

---

S. Stiller (✉)
Institut für Mathematische Optimierung, Technische Universität Braunschweig,
Pockelsstraße 14, 38106 Braunschweig, Germany
e-mail: sebastian.stiller@tu-braunschweig.de

entire group. If the rth group is offered the service, but the remaining capacity is less than this group's size, no member of the group will go to the show, and the entire remaining capacity is passed on to number $r + 1$. If group number 1 is too large, a smaller group, holding number 2, can be lucky.

Knapsacks with Unknown Capacities

Those who eventually get tickets have to pay for them. The price for tickets is fixed for each group a priori, but can vary substantially for different groups. The service provider seeks to maximize the revenue from the tickets of the admitted groups. Thus, from his vantage point there is an underlying knapsack problem.

In fact, it is a knapsack with unknown capacity: Each group is a knapsack item $r \in R = (1, \ldots, n)$ for which size ℓ_r and value v_r, the price of the tickets, are known when constructing the waiting list. A waiting list L is a permutation of the items R. The capacity B of the knapsack is not known when the waiting list L is constructed. Note that items do not arrive online; we assume they are known all at once when the provider orders the list.

To focus this presentation on the essential insights (and without loss of generality) let us assume that all items have different size, value and density $d_r := v_r/\ell_r$.

A Robust Objective

Usually, every waiting list yields a suboptimal knapsack solution for some capacity B. We seek a waiting list that for each capacity yields a solution whose value is close to the optimal value of that capacity. Formally, our objective reads as follows:

$$\min_{L \in S(R)} \max_{B \leq \sum_{r \in R} \ell_r} \frac{OPT(B)}{L(B)},$$

where $S(R)$ is the symmetric group over R and $L(B)$ is the value of items packed according to list L in case of capacity B. Further, $OPT(B)$ is the value of an optimal solution to the standard combinatorial knapsack problem with item set R and capacity B. For a waiting list L, we call $\max_B \frac{OPT(B)}{L(B)}$ its *robustness factor*. The robustness factor of an algorithm to construct waiting lists is the worst robustness factor over all instances.

Stubborn or Eager to Learn?

Waiting lists are constructed without knowing the capacity. At a closer look, full knowledge of the capacity is not even required when turning a waiting list into a packing. Instead, one can think of this procedure as an algorithm that probes the items on the waiting list one after the other. In case a probed item still fits within the capacity, the algorithm packs it and proceeds to the next item. In case it does not fit, the algorithm discards it and proceeds to the next item.

Slightly generalizing waiting lists, we consider any algorithm that uses probing and packing steps. In a probing step, the algorithm inquires whether a single probed item still fits with the current packing. In a packing step, the algorithm packs one item irrevocably. Throughout this paper, we call any algorithm a *packing policy* if it only uses these two types of steps.

Waiting lists are a special case of packing policies in two respects: First, if a probing is affirmative for an item, waiting lists immediately pack this item. Second, waiting lists follow an order fixed in advance for probing the items. Thus, waiting lists to some extend refuse to learn about the capacity. Surprisingly, in some worst-case instances this stubbornness means no loss for waiting lists.

What Shall We Hope For?

Can we construct a waiting list that provides a good packing for every capacity?

We devise a relatively simple way to order any finite set of items R, such that $\max_B \frac{OPT(B)}{L(B)}$ is never larger than 2, i.e., its robustness factor is upper bounded by 2.

The second result we describe here is a family of instances for which—in the limit, as the size of the instances goes to infinity—every packing policy for some capacity achieves at most half the value of an optimal knapsack solution.

Combining the two results, in worst-case terms the waiting lists we construct are best possible among all packing policies.

2 Algorithm

Ignoring Safety Regulations

Our objective compares for each capacity B the packing induced by the waiting list L to $OPT(B)$, an optimal knapsack solution for that capacity. A set of optimal knapsack solutions for all capacities does not provide for a structure easily amenable to analysis. Neither does a set of solutions we get from the well-known approximation schemes. Optimal or near-optimal solutions react too sensitively to changes in capacity.

To attain our positive result we keep things simple. We replace the set of optimal solutions by a nicely structured set of 2-approximate solutions. Instead of trying to be as close as possible to (near-)optimal solutions, we try to be at least as good as those 2-approximations. In light of our second result this is perfectly enough. Such a nicely structured set of 2-approximative solutions readily falls out of a folklore 2-approximation algorithm for the standard deterministic knapsack problem:

1. Order items $r \in R$ (with $\ell_r \leq B$) by decreasing density $d_r := v_r/\ell_r$.
2. Pack the better of the following two sets:
   - The longest prefix of the density ordering that fits within capacity B.
   - The first item after this prefix.

Imagine the organizers of the Stones tour ignore safety limits and admit the prefix *and* the next item to the concert. This way they would at least make as much profit as an optimal knapsack solution (no feasible solution can be denser on average over the densest B units of what is packed). Taking the better of two parts is at least half as good. Thus, the folklore algorithm is a 2-approximation.

If for a set of items R there is a capacity B for which the folklore algorithm chooses the next item over the prefix, we call that item the *swap item* for capacity B—because it swaps with the prefix. Formally, $s \in R$ is a swap item for capacity B, if $\ell_s \leq B$ and

$$v_s > \sum_{\substack{r \in R: \\ \ell_r \leq B, d_r > d_s}} v_r .$$

Folklore with Attention to Detail

The idea of the folklore algorithm seems well suited for the construction of waiting lists. In the remainder we describe an algorithm to construct waiting lists that imitate the behavior of the folklore algorithm for every capacity. For instances without swap items this is easy. Ordering the waiting list by density gives a waiting list L perfectly imitating the behavior of the folkore algorithm.

The natural idea to imitate swapping is to put the swap items in front. But, whether an item is a swap item depends on the capacity of the knapsack—which unfortunately is unknown. This is due to a seemingly trivial detail of the folklore algorithm. A very first step of the algorithm discards all items that are larger than capacity B. This changes which item is a swap item. For example, an item r larger and denser than all other items inhibits any item to be a swap item. But, there could well be swap items for capacities smaller than the size of item r. Thus, without the trivial deletion step the information the folklore algorithm builds upon is destroyed.

The following algorithm combines the idea of density ordering and putting swap items in front in just the right way.

How Shall the Provider Order the Waiting List?

We think of waiting lists as ordered vertically and let items drown into our lists in an unnatural way: the less dense they are, the deeper they drown. The *swap-and-drown-algorithm* works as follows:

1. Flag all items that are the swap item for some capacity.
2. Sort all items ascending by size.
3. Insert the items r into L in order of increasing size by the following rules:

    - If r is flagged as a swap item, put it on top of the list constructed so far.
    - Else let r drown into the list from the top until it stops right above the first item of lower density.

For a first intuition why the lists constructed by the swap-and-drown algorithm are good, observe the following: As swap-and-drown inserts the smaller items first, each swap item is above all smaller items, but not higher than larger and denser ones. To get a precise understanding, we start with two lemmata regarding the structure of these lists.

3 Analysis

Some Structure of the Swap-and-Drown List

We prepare the lemmata by observations that are immediate from the definition of the algorithm:

1. Swap items decrease in value.
   Due to the initial ordering by size, the swap items become smaller as we go down the list. Therefore, they also become less valuable: a smaller swap item is either part of what the larger one swaps, or less dense than the larger one.
2. Swap items are less valuable than the items before.
   An item is larger than any swap item lower in the list, because being inserted later is the only way to end up higher than a swap item. Due to the drowning rule a non-swap item is also denser than the next lower swap item in the list, and thus more valuable.

Combining these observation we have the following lemma.

Lemma 1 *Let L be the list constructed by the swap-and-drown-algorithm, and let s be a swap item. Each item r higher in the list than s is of higher value than s, i.e., $v_r > v_s$.*

Assume again s is a swap item. Together the items of size at most ℓ_s and density at least d_s are worth no more than s—because s swaps them. An item r larger than s can only get below s in our list, if it sinks past s, i.e., $d_r < d_s$. Thus, in our list the items below s with higher density than d_s are altogether worth no more than s. This statement will be instrumental, so we highlight it as our second lemma.

Lemma 2 *Let L be the list constructed by our algorithm, and let s be a swap item. Define R_s as the set of items r lower in the list L than s and with higher density, i.e., $d_r > d_s$. Then we have $v_s \geq \sum_{r \in R_s} v_r$.*

The Imitation Proof

We now show that the list constructed by the swap-and-drown algorithm imitates the folklore algorithm for every capacity. Thus, for all capacities, it is at most a factor of 2 away from an optimal knapsack solution. In short:

Theorem 3 *The swap-and-drown-algorithm has a robustness factor of at most 2.*

We look separately at the cases where the folklore algorithm chooses a swap item or a prefix. In the first case, Theorem 3 follows directly from Lemma 1: For a capacity where the folklore chooses swap item s, the list will pack s or at least one item before s. But, all items before s are at least as valuable as s. The prefix case needs a little more work.

The First Perpetrator

For a capacity where the folklore algorithm packs the prefix, the packing of the swap-and-drown waiting list has packed at least as much value as the prefix as soon as the list has packed the first item not used in the solution of the folkore algorithm. We will prove this now.

Let us pack according to the list, one by one, until the first item r which the list packs, but the folklore algorithm does not include—the first perpetrator. So, for this part of the proof of Theorem 3 we stop the list packing after r. We show that R'_r, the set of items the folklore algorithm packed, but we did not pack, is worth no more than r.

First, as the folklore algorithm packs the prefix, but did not include r, all items in R'_r are denser than r. Second, as the waiting list does not pack anything from R'_r before r, but r is the first perpetrator, the items in R'_r lie lower in the list than r.

Items denser and lower in the list than r can only exist if r is a swap item or there is a swap item between r and these items. Let s be the first swap item below r that is higher than any item in R'_r. Possibly, we have $r = s$. By the third step of the swap-and-drown algorithm, the density of r is not smaller than that of s.

Let us put the pieces together: First, the item r is at least as valuable as s, by Lemma 1. Second, a swap item s by Lemma 2 is more valuable than all items in R_s, i.e., those denser and deeper in the list than s. And third, the items in R'_r lie lower than s. They are denser than s, as r is as least as dense as s, so we have $R'_r \subseteq R_s$. We can conclude that r is at least as valuable as all items in R'_r together, and our packing is at least as good as the folklore packing.

Whatever You Pick, It Is Wrong!

A robustness factor of 2 is best possible even for general packing policies. The instances we use to show this allow for the following: Whenever a packing policy packs its first item, we can choose a capacity for which packing this item already means failure—by a factor of 2. Further, all capacities we use, are at least as large as each item. Thus, before the first packing step no probing step can teach the policy which of the relevant capacities we choose. But, after the first packing step the policy is already doomed.

In fact, we construct a family of instances with n items such that for any constant factor strictly less than 2, the instances larger than a certain n' inhibit such a factor.

These instances have the following two features.

1. The value of the items increase so slowly from item 1 to n that two consecutive items have roughly twice the value of the next item after them. Eventually the increase will be enough, so that the last item is almost twice as valuable as the first or the second item.
2. The sizes also increase from 1 to n, producing the following effect: When a policy packs an item $i > 3$, the capacity needed for the two items before i is enough for i, but for no second item.

Of course, this trick cannot work if a policy picks one of the two smallest items. For this case we pick the size of the last and largest item as capacity B, which is just too small to pack either item 1 or 2 with any second item.

To implement these features in a set of n items, we define the value of the ith item as $c_i = 1 + 1/n$ and its size as $F_i + F_n - 1$, where F_i is the ith Fibonacci number. If the policy packs one of the first two items, we choose as capacity the size of the last item, and the policy is off by a factor of $2 - 4/n$. If it packs another item first, we choose as capacity the sum of the two smaller items' sizes. In this case the optimal knapsack is a factor of $2 - 3/n$ better than the policy's choice. Thus, we get a matching lower bound for the robustness factor:

Theorem 4 *No packing policy can attain a robustness factor of $2 - \varepsilon$, for any constant $\varepsilon > 0$.*

4 Notes on the Literature

The knapsack problem is one of the most basic combinatorial optimization problems. In 1972, Karp [1] proved that it is NP-hard. In fact, the problem is weakly NP-hard. Early dynamic programming algorithms (e.g. [2]) that solve the problem in pseudopolynomial time were followed in 1975 by the first fully polynomial approximation scheme for the knapsack problem, due to Ibarra and Kim [3]. The knapsack is a role model for a class of problems for which a pseudopolynomial dynamic program can be turned into an FPTAS [4]. The book of Kellerer, Pferschy and Pisinger [5] provides an excellent, comprehensive, and updated overview of the knapsack problem. Even therein no reference is given for the folklore algorithm. It is simply folklore.

This paper is an easy-listening version with slightly rearranged proofs for some of the key results from a joint paper [6] with Yann Disser, Max Klimm, and Nicole Megow. In the beginning of the swap-and-drown algorithm one has to identify the swap items. A swap item s for some capacity B is also a swap item for the capacity equal to its own size ℓ_s. Thus, the algorithm is obviously in $O(n^2)$ for $n = |R|$. As shown in [6] the identification of all swap items and the entire swap-and-drown algorithm can be implemented to run in $O(n \log n)$ time. That paper, in addition, shows a similar, tight result for the case of unit density, where the factor of 2 is replaced by the golden ratio. It also addresses some of the pertaining hardness questions. For example, it is coNP-hard to decide, given an instance, whether a packing policy with some prescribed robustness factor α exists.

References

1. Karp, R.M.: Reducibility among combinatorial problems. In: Miller, R.E., Thatcher, J.W. (eds.) Complexity of Computer Computations, pp. 85–103. Plenum, New York (1972)
2. Bellmann, R.: Notes on the theory of dynamic programming IV—Maximization over discrete sets. Nav. Res. Q. **3**, 67–70 (1956)
3. Ibarra, O.H., Kim, C.E.: Fast approximation algorithms for the knapsack and the sum of subsets problem. J. ACM **22**(4), 463–468 (1975)
4. Woeginger, G.: When does a dynamic programming formulation guarantee the existence of a fully polynomial time approximation scheme (FPTAS)? INFORMS J. Comput. **12**(1), 57–74 (2000)
5. Kellerer, H., Pferschy, U., Pisinger, D.: Knapsack Problems. Springer, Berlin (2004)
6. Klimm, M., Disser, Y., Megow, N., Stiller, S.: Packing a knapsack of unknown capacity. In: Proceedings of the 31st International Symposium on Theoretical Aspects of Computer Science (STACS 2014), Leibniz International Proceedings in Informatics (LIPIcs), vol. 25, pp. 276–287. Schloss Dagstuhl–Leibniz-Zentrum für Informatik (2014)

Printed by Printforce, the Netherlands